The TI-83 Manual

Kathleen McLaughlin
Dorothy Wakefield

Elementary Statistics
PICTURING THE WORLD

LARSON ∎ FARBER

PRENTICE HALL, Upper Saddle River, NJ 07458

Acquisitions Editor: *Kathy Boothby Sestak*
Supplement Editor: *Joanne Wendelken*
Special Projects Manager: *Barbara A. Murray*
Production Editor: *Shea Oakley*
Manufacturing Buyer: *Alan Fischer*
Supplement Cover Manager: *Paul Gourhan*
Supplement Cover Designer: *Liz Nemeth*

© 2000 by **PRENTICE-HALL, INC.**
Upper Saddle River, NJ 07458

Printed in the United States of America

10 9 8 7 6 5 4 3

ISBN 0-13-015221-8

Prentice-Hall International (UK) Limited, *London*
Prentice-Hall of Australia Pty. Limited, *Sydney*
Prentice-Hall Canada, Inc., *Toronto*
Prentice-Hall Hispanoamericana, S.A., *Mexico*
Prentice-Hall of India Private Limited, *New Delhi*
Prentice-Hall (Singapore) Pte. Ltd.
Prentice-Hall of Japan, Inc., *Tokyo*
Editora Prentice-Hall do Brasil, Ltda., *Rio de Janeiro*

▶ Introduction

The TI-83 Graphing Calculator Manual is one of a series of companion technology manuals that provide hands-on technology assistance to users of Larson/Farber *Elementary Statistics: Picturing the World.*

Detailed instructions for working selected examples, exercises, and Technology Labs from *Elementary Statistics: Picturing the World* are provided in this manual. To make the correlation with the text as seamless as possible, the table of contents includes page references for both the Larson/Farber text and this manual.

▶ Contents:

Getting Started with the TI-83 Graphing Calculator

▶ Overview

The TI-83 graphing calculator has a variety of useful functions for doing statistical calculations and for creating statistical plots. Before you begin using the TI-83, spend a few minutes becoming familiar with its basic operations. First, notice the different colored keys on the calculator. The gray keys are the number keys. The blue keys along the right side of the keyboard are the common math functions. The blue keys across the top set up and display graphs. The primary function of each key is printed in white on the key. For example, when you press STAT, the STAT MENU is displayed.

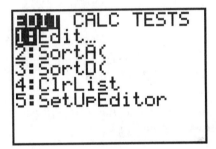

The secondary function of each key is printed in yellow above the key. When you press the 2nd key, the function printed in yellow above the key becomes active and the cursor changes from a solid rectangle to an ↑ (up-arrow). For example, when you press 2nd and the $\boxed{x^2}$ key, the $\sqrt{}$ function is activated. In this manual, all 2nd functions are highlighted in gray. For example, to use the LIST function, found above the STAT key, the notation used in this manual is 2nd [LIST]. The LIST MENU will then be activated and displayed on the screen.

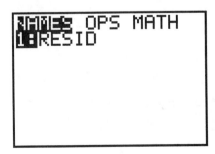

The alpha function of each key is printed in green above the key. When you press the green ALPHA key, the function printed in green above the key is activated and the cursor changes from a solid rectangle to A.

In this manual you will find detailed explanations of the different statistical functions. These explanations will accompany selected examples from your textbook.

▶ Getting Started

To operate the calculator, press ON in the lower left corner of the calculator. Begin each example with a blank screen, with a rectangular cursor flashing in the upper left corner. If you turn on your calculator and you do not have a blank screen, press the CLEAR key. You may have to press CLEAR a second time in order to clear the screen. If using the CLEAR key does not clear the screen, you can push 2nd [QUIT] (Note: QUIT is found above the MODE key.)

▶ Helpful Hints

To adjust the display contrast, push and release the 2nd key. Then push and hold the blue up arrow ▲ to darken or the blue down arrow ▼ to lighten.

The calculator has an automatic turn off that will turn the calculator off if it has been idle for several minutes. To restart, simply press the ON key.

There are several different graphing techniques available on the TI-83. If you inadvertently leave a graph on and attempt to use a different graphing function, your graph display may be cluttered with extraneous graphs, or you may get an ERROR message on the screen.

There are several items that you should check before graphing anything. First, press the Y= key and clear all the Y-variables. The screen should look like the following display:

```
Plot1  Plot2  Plot3
\Y1=
\Y2=
\Y3=
\Y4=
\Y5=
\Y6=
\Y7=
```

If there are any functions stored in the Y-variables, simply move the cursor to the line that isn't clear and press CLEAR ENTER.

Next, press 2nd [STAT PLOT] and check to make sure that all the STAT PLOTS are turned OFF.

```
STAT PLOTS
1:Plot1...Off
    L1    L2    □
2:Plot2...Off
    L1    L2    □
3:Plot3...Off
    L1    L2    □
4↓PlotsOff
```

If you notice that a Plot is turned ON, select the Plot, press ENTER and move the cursor to OFF and press ENTER. Press 2nd [QUIT] to return to the home screen.

Introduction to Statistics

CHAPTER
1

▶ Technology Lab (pg. 22-23) Generating Random Numbers

To generate a random sample of integers, press **MATH** and the Math Menu will appear.

```
MATH NUM CPX PRB
1▶Frac
2:▶Dec
3:3
4:3√(
5:×√
6:fMin(
7↓fMax(
```

Use the blue right arrow key, 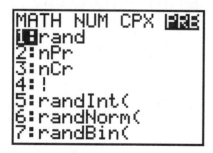, to move the cursor to highlight **PRB**. The Probability Menu will appear.

```
MATH NUM CPX PRB
1▶rand
2:nPr
3:nCr
4:!
5:randInt(
6:randNorm(
7:randBin(
```

Select **5:RandInt(** by using the blue down arrow key, ▼ , to highlight it and pressing **ENTER** or by pressing the 5 key. **RandInt(** should appear on the screen. This function requires three values: the starting integer, followed by a comma (the comma is found on the black key above the 7 key), the ending integer, followed by a comma and the number of values you want to generate. Close the parentheses and press **ENTER**. (Note: It is optional to close the parenthesis at the end of the command.)

For an example, suppose you want to generate 15 values from the integers ranging from 1 to 50. The command is **randInt(1,50,15)**.

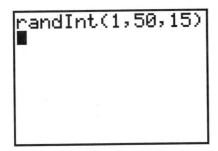

Press **ENTER** and a partial display of the 15 random integers should appear on your screen. (Note: your numbers will be different from the ones you see here.)

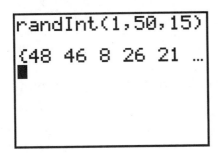

Use the right arrow to scroll through your 15 items. You might find that you have some duplicate values. The TI-83 uses a method called "sampling with replacement" to generate random numbers. This means that it is possible to select the same integer twice.

In the example in the text, you are asked to select a random sample of 15 cars from the 167 cars that are assembled at an auto plant. One way to choose the sample is to number the cars from 1 to 167 and randomly select 15 different cars. This sampling process is to be "without replacement." Since the TI-83 samples "with replacement", the best way to obtain 15 different cars is to generate more than 15 random integers, and to discard any duplicates. To be safe, you should generate 20 random integers.

```
randInt(1,167,20
)
```

```
randInt(1,167,20
)
{146 119 40 81 ...
■
```

Use the right arrow to scroll to the right to see the rest of the list and write down the first 15 distinct values.

Exercises

1. You would like to sample 8 distinct accounts. The TI-83 samples "with replacement" so you may have duplicates in your sample. To obtain 8 distinct values, try selecting 10 items. Press **MATH**, highlight **PRB** and select **5:RandInt(**. Press **ENTER** and type in **1,74,10).** Press **ENTER** and the random integers will appear on the screen. Use the right arrow to scroll through the output and choose the first eight distinct integers. (Note: If you don't have 8 distinct integers, generate 10 more integers and pick as many new integers as you need from this group to complete your sample of 8).

2. Because you want 20 distinct batteries, generate 25 random integers using **randInt(1,200,25).** Use the right arrow to scroll through the list and select the first 20 distinct integers.

3. This example does not require distinct digits, so duplicates are allowed. You can select three random samples of size n=5 by using **randInt** to create each sample and then store each sample. Press **MATH**, highlight **PRB** and select **5:RandInt(** by scrolling down through the list and highlighting **5:RandInt(** and pressing **ENTER** or by highlighting **PRB** and pressing **5**. Type in the starting digit: 0; the ending digit: 9, and the number of selections: 5.

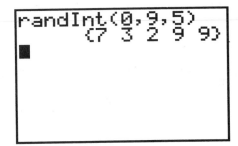

Add the digits and divide by 5 to get the average.
Repeat this process two more times.

Note: To calculate the population average, add the digits 0 through 9 and
divide the answer by 10.

4. Use the same procedure as in Exercise 3 with a starting value of 0 and an
    ending value of 40 and a sample size of 7.

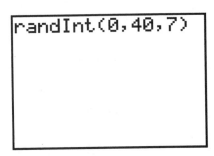

5. Use **randInt(1,6,60).** Store the results in L1 by pressing **STO** **2ⁿᵈ** **[L1]**
    ENTER. Then sort L1 by pressing **2ⁿᵈ** **[LIST]**, highlight **OPS** and select
    1:SortA(and press **2ⁿᵈ** **[L1]** **ENTER**. Press **2ⁿᵈ** **[L1]** **ENTER** and a
    partial list of the integers in L1 will appear on the screen. Use the right
    arrow to scroll through the list and count the number of 1's, 2's, 3's, ... and
    6's. Record the results in a table.

7. Use **randInt(0,1,100).** Store the results in L1. Press **2ⁿᵈ** **[LIST]** , highlight
    MATH and select **5:sum(** and press **2ⁿᵈ** **[L1]**. Close the parentheses and
    press **ENTER**.

The number in your output is the sum of L1. Since L1 consists of 0's and 1's, the sum is actually the number of tails in your random sample. The (number of heads) = 100 - (no. of tails).

CHAPTER
2

Descriptive Statistics

Section 2.1

▶ Example 7 (pg. 38) Construct a histogram using Internet data

To create this histogram, you must enter information into List1 (**L1**) and List 2 (**L2**). Refer to the frequency distribution on pg. 33 in your textbook. You will enter the midpoints into **L1** and the frequencies into **L2**. Press STAT and the Statistics Menu will appear.

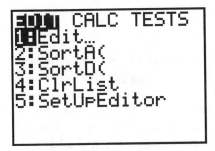

Press ENTER and lists **L1**, **L2** and **L3** will appear.

If the lists already contain data, you should clear them before beginning this example. Move your cursor so that the List name (**L1**, **L2**, or **L3**) of the list that contains data is highlighted.

```
 ▊1        L2        L3        1
  1         4       ------
  2         6
  3         8
 10
------

L1 ={1,2,3,10}
```

Press **CLEAR** **ENTER**. Repeat this process until all three lists are empty.

```
L1        L2        L3        1
▬▬▬▬▬       4       ------
           6
           8
          ------


L1(1)=
```

To enter the midpoints into **L1,** move your cursor so that it is positioned in the 1st position in **L1**. Type in the first midpoint, **12.5,** and press **ENTER** or use the blue down arrow. Enter the next midpoint, **24.5.** Continue this process until all 7 midpoints are entered into **L1**. Now use the blue up-arrow to scroll to the top of **L1**. As you scroll through the data, check it. If a data point is incorrect, simply move the cursor to highlight it and type in the correct value. When you have moved to the 1st value in **L1**, use the right arrow to move to the first position in **L2**. Enter the frequencies into **L2**.

```
L1        L2        L3        2
12.5      ▊         ------
24.5      10
36.5      13
48.5      8
60.5      5
72.5      6
84.5      2
L2(1)=6
```

To graph the histogram, press **2ⁿᵈ** [STAT PLOT] (located above the **Y=** key).

Select Plot1 by pressing **ENTER**.

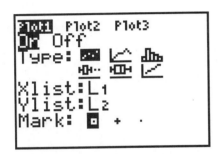

Notice that Plot1 is highlighted. On the next line, notice that the cursor is flashing on **ON** or **OFF**. Position the cursor on **ON** and press **ENTER** to select it. The next two lines on the screen show the different types of graphs. Move your cursor to the symbol for histogram (3rd item in the 1st line of **Type**) and press **ENTER**.

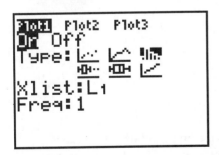

The next line is **Xlist**. Use the blue down arrow to move to this line. On this line, you tell the calculator where the data (the midpoints) are stored. In most graphing situations, the data are entered into **L1** so **L1** is the default option. Notice that the cursor is flashing on **L1**. Push **ENTER** to select **L1**. The last line is the frequency line. On this line, **1** is the default. The cursor should be flashing on **1**. Change **1** to **L2** by pressing 2nd **[L2]**.

To view a histogram of the data, press ZOOM.

There are several options in the Zoom Menu. Using the blue down arrow, scroll down to option 9, **ZoomStat,** and press ENTER. A histogram should appear on the screen.

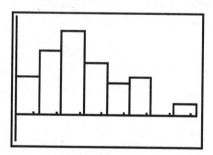

This histogram is not exactly the same as the one on pg. 38 of your textbook. You can adjust the histogram so that it does look exactly like the one in your text. Press Window and set **Xmin** to 12.5, **Xmax** to 96.5 (this one extra midpoint is needed to complete the picture), and **Xscl** equal to 12, which is the difference between successive midpoints in the frequency distribution.

```
WINDOW
 Xmin=12.5
 Xmax=96.5
 Xscl=12
 Ymin=-3.90897
 Ymax=15.21
 Yscl=.1
 Xres=1
```

Press **GRAPH**.

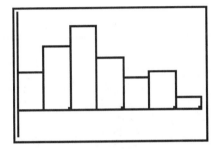

Notice the blue **TRACE** key. If you press it, a flashing cursor, ✳, will appear at the top of the 1st bar of the histogram.

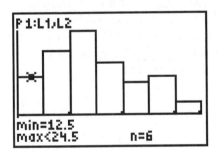

The value of the smallest midpoint is displayed as 12.5 and the number of data points in that bar is displayed as n = 6. Use the blue right arrow to move through each of the bars.

Now that you have completed this example, turn Plot1 **OFF**. Using 2nd [STAT PLOT], select Plot1 by pressing **ENTER** and highlighting **OFF**. Press **ENTER** and 2nd [QUIT]. (Note: Turning Plot1 **OFF** is optional. You can leave it ON but leaving it ON will effect other graphing operations of the calculator.)

◀

▶ Exercise 17 (pg. 41) Construct a frequency histogram using 6
classes

Push **STAT** and **ENTER** to select **1:Edit**. Highlight **L1** at the top of the first list
and press **CLEAR** and **ENTER** to clear the data in **L1**. You can also clear **L2**
although you will not be using **L2** in this example. Enter the data into **L1**. Scroll
through your completed list and verify each entry.

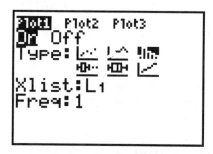

To set up the histogram, push **2ⁿᵈ** [STAT PLOT] and **ENTER** to select **Plot 1**.
Turn ON **Plot 1**, set **Type** to **Histogram**, set **Xlist** to **L1**. In this example, you
must set **Freq** to **1**. If the frequency is set on **L2** move the cursor so that it is
flashing on **L2** and press **CLEAR** . The cursor is now in ALPHA mode (notice
that there is an "A" flashing in the cursor). Push the **ALPHA** key and the cursor
should return to a solid flashing square. Type in the number **1.**

Press **ZOOM** and **9:ZoomStat** and press **ENTER** and a histogram will appear on
the screen. Press **Window** to adjust the Graph Window. Set **Xmin** equal to 1000
(the smallest data value) and **Xmax** equal to 7600 (the endpoint of the largest
class). To set the scale so that you will have 6 classes, calculate (**Xmax -
Xmin**)/6. Round this number to the next highest hundred (1100) and use this for
Xscl.

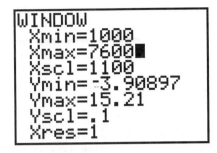

Press GRAPH and the histogram should appear.

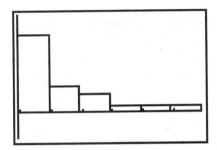

You can press TRACE and scroll through the bars of the histogram.
Min and Max values for each bar will appear along with the number of data
points in each class.

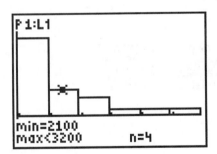

Notice that there are 4 data points in the second class (2100 to 3200)

Section 2.2

▶ Example 5 (pg. 48) Constructing a Pareto Chart

To construct a Pareto chart to represent the causes of inventory shrinkage, you must enter the causes of inventory shrinkage. Enter labels for the specific causes into **L1** and the costs into **L2**. Since the bars are positioned in descending order in a Pareto chart, the labels in **L1** will represent the causes of inventory shrinkage from the most costly to the least costly. Press **STAT** and select **1:Edit** and press **ENTER**. Highlight the name "**L1**" and press **CLEAR** and **ENTER**. Enter the numbers 1,2,3 and 4 into **L1**. These numbers represent the four causes of inventory shrinkage: 1 = employee theft, 2 = shoplifting, 3 = administrative error and 4 = vendor fraud. Move your cursor to highlight "**L2**" and press **CLEAR** and **ENTER**. Enter the values 15.6, 14.7, 7.8 and 2.9.

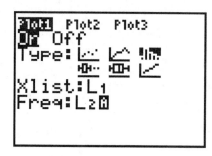

To draw the Pareto chart, press **2nd** [STAT PLOT] . Press **ENTER** and set up **Plot 1**. Highlight **On** and press **ENTER**. Highlight the histogram icon for **Type** and press **ENTER**. Set **Xlist** to **L1** and **Freq** to **L2**.

Press **ZOOM** and select **9:ZoomStat** and **ENTER**.

To adjust the picture so that the bars are connected press **Window**. Set **Xmax** =
5 and **Xscl** = 1, and press **GRAPH**.

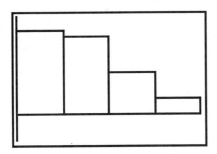

If you press **TRACE**, notice min=1, max < 2 and n = 15.6. min =1 is the
minimum value of the 1st bar. In this example, the first bar represents the No. 1
cause of inventory shrinkage. N = 15.6 is the frequency (in millions of dollars)
for cause #1. You can use the right arrow key to scroll through the remaining

causes.

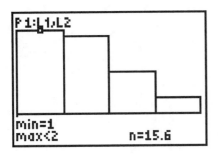

▶ Example 7 (pg. 50) Constructing a Time Series Chart

Press **STAT** and select **1:Edit** from the **Edit Menu**. Clear **L1** and **L2**. Enter the "years" into **L1** and the "subscribers" into **L2**.

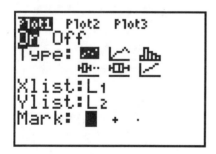

To construct the time series chart, press **2nd** [STAT PLOT] and select **1:Plot 1** and **ENTER**. Turn ON **Plot 1**. Set the **Type** to **Scatterplot,** which is the 1st icon in the **Type** choices. For **Xlist** select **L1** and for **Ylist** select **L2**. Next, there are three different types of **Marks** that you can select for the graph. The first choice, a small square, is the best one to use.

Press **ZOOM** and scroll down to **9:ZoomStat** and press **ENTER** or simply press **9** and **ZoomStat** will automatically be selected. The graph should appear on the screen.

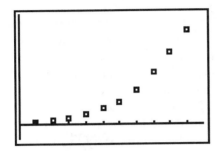

Use **TRACE** to scroll through the data values for each year. Notice for example, the number of subscribers is 24.1 million in 1994.

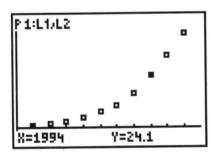

▶ Exercise 25 (pg. 53) Pareto Chart

Press **STAT** and select **1:Edit** from the **Edit Menu**. Clear the lists and enter the numbers 1 through 5 into **L1**. These numbers are labels for the five cities. (Note: 1= Denver, the city with the highest ultraviolet index, 2 = Atlanta, the city with the second highest ultraviolet index, etc.). Enter the ultraviolet indices in descending order into **L2**. Press **2ⁿᵈ [STAT PLOT]** and select **Plot 1** and press **ENTER**. Set the **Type** to **Histogram**. Set **Xlist** to **L1** and Freq to **L2**. Press **ZOOM** and press **9** for **ZoomStat.** Adjust the graph by pressing **Window** and setting **Xmax** = 6 and **Xscl** = 1. Press **GRAPH** to view the Pareto Chart.

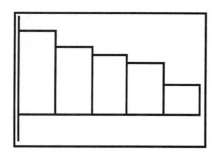

▶ Exercise 27 (pg. 54) Scatterplot

Press **STAT** and select **1:Edit** from the **Edit menu**. Clear the lists and enter "number of students per teacher" into **L1** and "average Teacher's salary" into **L2**. To construct the scatterplot, press **2ⁿᵈ** **[STAT PLOT]** and select **1:Plot 1** and **ENTER**. Turn ON **Plot 1**. Set the **Type** to **Scatterplot** which is the 1ˢᵗ icon in the **Type** choices. For **Xlist** select **L1** and for **Ylist** select **L2**. For **Marks** use the first choice, the small square.

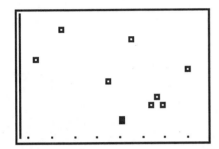

As you can see from the scatterplot, this data set has a large amount of scatter. You might notice a slight downward trend which suggests that average teacher's salaries decrease as the number of students per teacher increases.

◀

▶ Exercise 29 (pg. 54) Time Series Chart

Press **STAT** and select **1:Edit**. Enter the "Years" into **L1** and the "prices" into **L2**. Press **2ⁿᵈ** [STAT PLOT] and select 1: **Plot 1** and press **ENTER**. Turn **ON** **Plot 1** and select the **scatterplot** (1ˢᵗ icon) as **Type**. Set **Xlist** to **L1** and **Ylist** to **L2** and Mark to ■. Press **ZOOM** and press **9** for **ZoomStat** and view the graph.

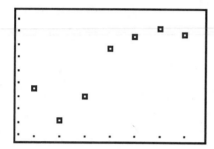

Section 2.3

▶ Example 6 (pg. 58) Comparing the mean, median, and mode

Press **STAT** and select **1:Edit**. Clear **L1** and enter the data into **L1**. Press **STAT** again and highlight **CALC** to view the Calc Menu.

```
EDIT CALC TESTS
1:1-Var Stats
2:2-Var Stats
3:Med-Med
4:LinReg(ax+b)
5:QuadReg
6:CubicReg
7↓QuartReg
```

Select **1:1-Var Stats** and press **ENTER ENTER** and the first page of the one variable statistics will appear.

```
1-Var Stats
 x̄=23.75
 Σx=475
 Σx²=13109
 Sx=9.808025715
 σx=9.559680957
↓n=20
```

The first item is the mean, $\bar{x} = 23.75$. Notice the down arrow in the bottom left corner of the screen. This indicates that more information follows this first page. Use the blue down arrow to scroll through this information. The third item you see on the second page is the median, Med = 21.5.

```
1-Var Stats
↑n=20
 minX=20
 Q₁=20
 Med=21.5
 Q₃=23
 maxX=65
■
```

The TI-83 does not calculate the mode but, since this data set is sorted, it is easy to see from the list of data that the mode is 20.

▶ Example 7 (pg. 59) Finding a Weighted Mean

Press **STAT** and select **1:Edit**. Clear **L1** and **L2**. Enter the scores into **L1** and the weights into **L2**.

```
L1      L2      L3      2
86      .5      ------
96      .15
82      .2
98      .1
100     .05

L2(6) =
```

Press **STAT** and highlight **CALC** to view the Calc Menu. Select **1:1-Var Stats**, press **ENTER** and press 2nd **[L1]** , 2nd **[L2]** . Press **ENTER**. (Note: You must place the comma between **L1** and **L2**).

```
1-Var Stats L₁,L
₂
```

Using **L1** and **L2** in the **1:1-Var Stats** calculation is necessary when calculating a weighted mean. In this example, the weighted mean is 88.6.

```
1-Var Stats
 x̄=88.6
 Σx=88.6
 Σx²=7885.6
 Sx=
 σx=5.969924623
↓n=1
```

◀

▶ Example 8 (pg. 60) Finding the Mean of a Frequency
Distribution

Press **STAT** and select **1:Edit**. Clear **L1** and **L2**. Enter the x-values into **L1**
and the frequencies into **L2**. Press **STAT**, highlight **CALC**, select **1:1-Var
Stats**, press **ENTER**. Next, press **2ⁿᵈ** **[L1]** **[,]** **2ⁿᵈ** **[L2]** **ENTER**.

After you press **ENTER**, the sample statistics will appear on the screen. The
mean of the frequency distribution described in columns **L1** and **L2** is 41.78. In
this example you do not have the actual data. What you have is the frequency
distribution of the data summarized into categories. The mean of this frequency
distribution is an approximation of the mean of the actual data.

```
1-Var Stats
 x̄=41.78
 Σx=2089
 Σx²=107196.5
 Sx=20.16163259
 σx=19.95899797
↓n=50
```

▶ Exercise 17 (pg. 63) Finding the Mean, Median and Mode

Enter the data into **L1**. Press **STAT** and select **1:1-Var Stats** from the Calc
Menu. Press **ENTER** **ENTER**.

```
1-Var Stats
 x̄=97
 Σx=2716
 Σx²=263903.82
 Sx=4.090730262
 σx=4.017017373
↓n=28
```

The first item in the output screen is the mean, 97. Scroll down through the
output and find the median, Med = 97.2.

```
1-Var Stats
↑n=28
 minX=87.5
 Q₁=95.05
 Med=97.2
 Q₃=100.1
 maxX=103.1
■
```

Although the TI-83 does not calculate the mode, you can use the SORT feature to
order the data. You can then scroll through the data to see if the data set contains a
mode. To sort the data, press **2ⁿᵈ** **[LIST]**   (Note: **List** is found above the **STAT**
key). Move the cursor to highlight **OPS** and select 1:**SortA(** and **ENTER**.

```
NAMES OPS MATH
1 SortA(
2:SortD(
3:dim(
4:Fill(
5:seq(
6:cumSum(
7↓ΔList(
```

To sort **L1** in ascending order, press **2ⁿᵈ** **[L1]** , close the parentheses and press **ENTER**.

```
SortA(L₁)
              Done
■
```

To view the data, press **STAT** and select **1:Edit** and press **ENTER**. Use the down arrow to scroll through **L1** to see if the data has a mode. In this example, there is no mode because there is no individual data value that occurs more often than all the others. (Most of the data values occur only once and there are four data values that occur twice.)

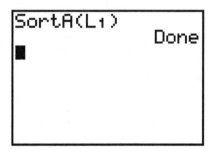

```
L1      L2      L3      2
87.5    ▮▮▮▮▮   ------
89.2
90.5
90.6
94.2
94.8
94.8
L2(1)=
```

▶ Exercise 35 (pg. 65)   Finding the Mean of Grouped Data

Press **STAT** and select **1:Edit**. Clear **L1** and **L2**. Enter the midpoints of each Age group into **L1** and the frequencies into **L2**.

L1	▨	L3	2
4.5	57	------	
14.5	68		
24.5	36		
34.5	55		
44.5	71		
54.5	44		
64.5	36		

L2 = {57,68,36,55...

Press **STAT** and highlight **CALC** to view the Calc Menu. Select **1:1-Var Stats**, press **ENTER** and type in **L1** , **L2.** The mean of this grouped data is 35.01.

```
1-Var Stats
 x̄=35.01413882
 Σx=13620.5
 Σx²=658407.25
 Sx=21.62813485
 σx=21.6003173
↓n=389
```

◀

▶ Exercise 39 (pg. 66) Construct a frequency histogram

Press **STAT** and select **1:Edit**. Clear **L1** and enter the data into **L1**. Press **2ⁿᵈ**
[STAT PLOT] and select **1:Plot 1** and press **ENTER**. Turn **ON** Plot 1 and set
the **Type** to **Histogram** and press **ENTER**. Move the cursor to **Xlist** and set this
to **L1**. Move the cursor to **Freq** and set it equal to **1.** (Note: The cursor may be
in ALPHA mode with a flashing Ⓐ . Press **ALPHA** to return to the solid
rectangular cursor and type in **1** .

Press **ZOOM** and **9** to select **ZoomStat.** The histogram that is displayed
has 7 classes. To change to 6 classes, press **Window**. Notice that **Xmin** =
62 and **Xmax** = 79.5. To determine a value for **Xscl**, you must calculate
(**Xmax - Xmin**)/6. Press **2ⁿᵈ** **[QUIT]** to get out of the Window Menu.
(Note: QUIT is found above the MODE key). Calculate (79.5 - 62)/6 and
round the resulting value, 2.9, to 3. Press **Window** and set **Xscl** = 3. Set
Xmax = 80. Press **GRAPH** and view the histogram with 6 classes. As
you can see from the graph, the histogram appears to be symmetric.

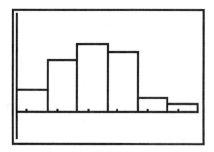

▶ Exercise 45 (pg. 67)   Data Analysis

Press **STAT** and select **1:Edit**. Clear **L1** and **L2** and then enter the data. Press **STAT** and highlight **CALC**. Select **1:1-Var Stats** and press **ENTER** **ENTER** to get values for the mean and median.

Since the TI-83 does not do a stem and leaf display, you can use a histogram to get a picture of the data. Press 2nd [STAT PLOT] , select **1:Plot 1** and press **ENTER**. Turn ON **Plot 1**. Set **Type** to **Histogram**. The **Xlist** is **L1** and the **Freq** is **1.** Press **ZOOM** and **9** to select **ZoomStat.**

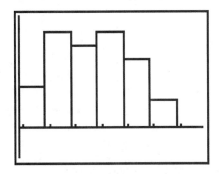

Section 2.4

▶ Example 5 (pg. 72) Finding the Standard Deviation

Press **STAT** and select **1:Edit**. Clear **L1** and enter the data into **L1**. Press **STAT** and highlight **CALC** to display the Calc Menu. Select **1: 1-Var Stats** and press **ENTER ENTER**. The sample standard deviation is Sx, 5.49224379.

```
1-Var Stats
 x̄=23.46875
 Σx=563.25
 Σx²=13912.5625
 Sx=5.49224379
 σx=5.376604654
↓n=24
```

▶ Example 9 (pg. 76) Standard Deviation of Grouped Data

Press **STAT** and select **1:Edit**. Clear **L1** and **L2**. Enter the x-values into **L1** and the frequencies into **L2**. Press **STAT** and highlight **CALC** to select the Calc Menu. Select **1:1-Var Stats** and press **ENTER**. Type in **L1** ⎵ **L2** and press **ENTER**. From the statistics displayed, the mean is 1.82 and the sample standard deviation is 1.722.

```
1-Var Stats
 x̄=1.82
 Σx=91
 Σx²=311
 Sx=1.722480414
 σx=1.705168613
↓n=50
```

▶ Exercise 17 (pg. 79) Comparing Two Datasets

Press **STAT** and select **1:Edit**. Clear **L1** and **L2**. Enter the Los Angeles Data into **L1** and the Long Beach Data into **L2**. Press **STAT** and highlight **CALC**. Select **1:1-Var Stats** and press **ENTER** **ENTER**. The sample standard deviation for the Los Angeles data is Sx = 6.111.

```
1-Var Stats
 x̄=26.25555556
 Σx=236.3
 Σx²=6502.97
 Sx=6.111282826
 σx=5.761772704
↓n=9
■
```

To find the range, scroll down through the display and find **minX = 18.3.** Continue scrolling through the display and find **maxX = 35.9.** To calculate the range, simply type in **35.9 - 18.3** and press **ENTER**. The range = 17.6.

```
minX=18.3
Q₁=20.55
Med=26.1
Q₃=31.85
maxX=35.9
35.9-18.3
            17.6
```

To find the variance, you must square the standard deviation. Type in 6.111 and press the x^2 key. The variance is 37.344.

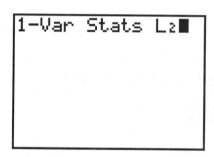

```
Med=26.1
Q3=31.85
maxX=35.9
35.9-18.3
              17.6
6.111²
        37.344321
```

For the Long Beach Data, press **STAT** , highlight **CALC**, select **1:1-Var Stats**, press **ENTER** and press 2nd **L2**.

```
1-Var Stats L2█
```

Press **ENTER**. The sample standard deviation for the Long Beach Data is Sx = 2.952.

```
1-Var Stats
 x̄=22.87777778
 Σx=205.9
 Σx²=4780.25
 Sx=2.952023788
 σx=2.783194718
↓n=9
█
```

Use the same procedure as you used for the Los Angeles Data, to find the range and the variance for the Long Beach Data.

Section 2.5

▶ Example 2 (pg. 86) Finding Quartiles

Press **STAT** and select **1:Edit**. Clear **L1** and enter the data into **L1**. Press
STAT and highlight **CALC**. Select **1:1-Var Stats** and press **ENTER** **ENTER**.
Scroll down through the descriptive statistics. You will see the first quartile, **Q1**
= **21.5,** the second quartile (the median), **Med = 23** and the third quartile, **Q3 =**
28.

```
1-Var Stats
↑n=25
 minX=15
 Q₁=21.5
 Med=23
 Q₃=28
 maxX=30
```

◀

▶ Example 4 (pg. 88) Drawing a Box-and-Whisker-Plot

Press **STAT** and select **1:Edit**. Clear **L1** and enter the data from Example 1 on pg. 85 in your textbook. Press 2nd [STAT PLOT]. Select **1:Plot 1** and press **ENTER**. Turn On **Plot 1**. Using the right arrow (you can not use the down arrow to drop to the second line), scroll through the **Type** options and choose the second boxplot which is the middle entry in row 2 of the **TYPE** options. Press **ENTER**. Move to **Xlist** and type in **L1**. Press **ENTER** and move to **Freq**. Set **Freq** to **1**. If **Freq** is set on **L2**, press **CLEAR**, and press **ALPHA** to return the cursor to a flashing solid rectangle and type in **1**. Press **ZOOM** and **9** to select **ZoomStat**. The Boxplot will appear on your screen.

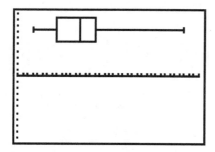

If you press **TRACE** and use the left and right arrow keys, you can display the five values that represent the five-number summary of the data.

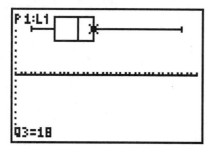

Notice that the trace cursor is on the right side of the box, which represents the third quartile. Also, at the bottom of the screen it is noted that Q3=18.

◀

▸ Example 11 (pg. 90) Quartiles and a Box-and-Whisker-Plot

Press **STAT** and select **1:Edit**. Clear **L1** and enter the data . Press **2ⁿᵈ** [STAT PLOT]. Select **1:Plot 1** and press **ENTER**. Turn On **Plot 1**. Using the right arrow, scroll through the **Type** options and choose the second boxplot which is the second entry in row 2 of the **TYPE** options. Press **ENTER**. Move to **Xlist** and type in **L1**. Press **ENTER** and move to **Freq**. Set **Freq** to **1**. If **Freq** is set on **L2**, press **CLEAR** , and press **ALPHA** to return the cursor to a flashing solid rectangle and type in **1**. Press **ZOOM** and **9** to select **ZoomStat.** The Boxplot will appear on your screen. If you press **TRACE** and use the right and left arrows, you can display **Q1, Med** and **Q3**. Notice Med=6 for this example.

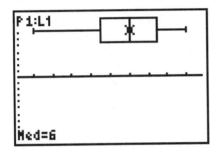

▶ Example 19 (pg. 91) Quartiles and a Box-and-Whisker-Plot

Press **STAT** and select **1:Edit**. Clear **L1** and enter the data. Press **2ⁿᵈ** [STAT PLOT]. Select **1:Plot 1** and press **ENTER**. Turn On **Plot 1**. Using the right arrow, scroll through the **Type** options and choose the second boxplot which is the second entry in row 2 of the **TYPE** options. Press **ENTER**. Move to **Xlist** and type in **L1**. Press **ENTER** and move to **Freq**. Set **Freq** to **1**. If **Freq** is set on **L2**, press **CLEAR** , and press **ALPHA** to return the cursor to a flashing solid rectangle and type in **1**. Press **ZOOM** and **9** to select **ZoomStat**. The Boxplot will appear on your screen. If you press **TRACE** and use the right and left arrows, you can display **Q1, Med** and **Q3**.

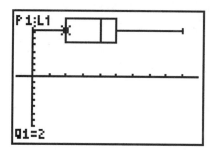

c. Using the quartiles and the Boxplot, you can make several statements about the data. Here are a few suggestions:

25% of the individuals in the sample watch less than 2 hours of television per day.

25% of the individuals in the sample watch more than 5 hours of television per day.

50% of the individuals in the sample watch between 2 and 5 hours of television per day.

Because the right whisker is somewhat longer than the left whisker, the data appears slightly skewed to the right.

◀

▶ Technology Lab (pg. 93) Monthly Milk Production

Exercises 1-2: Press **STAT** and select **1:Edit**. Clear **L1** and enter the data into
L1. Press **STAT** and highlight **CALC**. Select **1:1-Var Stats** and press **ENTER**
ENTER. The sample mean and sample standard deviation can be found in the
output display.

Exercises 3-4. You will use the histogram of the data to construct the frequency
distribution. To construct a histogram, press **2ⁿᵈ** [STAT PLOT]. Select **1: Plot
1** and **ENTER**. Turn ON **Plot 1** by highlighting **On** and pressing **ENTER**.
Scroll through the **Type** icons, highlight the **Histogram** and press **ENTER**.
Move to **Xlist** and type in **L1**. Move to the **Freq** entry and type in **1**. (Note: If
the Freq entry is **L2**, Press **CLEAR** , press **ALPHA** and enter **1**). To view a
histogram of the data, press **ZOOM** and **9**. To adjust the histogram, press
Window. Set **Xmin = 1000** and set **Xscl = 500**. (All other entries in the
Window Menu can remain unchanged). Press **GRAPH** and then press **TRACE**.

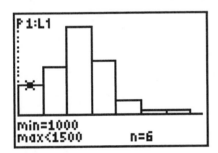

The minimum and maximum values of the first category are displayed. (**1000
and < 1500**). The midpoint of the first category is (1000+1499)/2, which is
1249.5. The frequency for the first category is **n = 6.** Trace through the
histogram and set up a frequency distribution for the data:

Category	Midpoint	Frequency

Exercise 5. Calculate the intervals $(\bar{x} \pm 1s)$ and $(\bar{x} \pm 2s)$. Press **2ⁿᵈ** [LIST].
Highlight **OPS** and select **1:SortA(** , press **ENTER**. Press **2ⁿᵈ** [L1] and close
the parentheses. Press **ENTER**. Press **STAT**, select **1:Edit** and press **ENTER**.
Scroll through the data in **L1** and count the number of data points that fall in the

intervals $(\bar{x} \pm 1s)$ and $(\bar{x} \pm 2s)$. To find the percentage of data points that lie in each of these intervals, calculate (no. of data points in each interval/ 50) * 100. Compare these percentages with the percentages stated in the Empirical rule.

Exercises 6 - 8. Using the frequency distribution you created, enter the midpoints into **L2** and the frequencies into **L3**. Press **STAT** , highlight **CALC**. Select **1:1-Var Stats** and press **ENTER**. Type in **L2** **,** **L3** and press **ENTER**. The mean and standard deviation for the frequency distribution will be displayed on the screen.

Probability

Section 3.1

▶ Law of Large Numbers (pg. 108)

You can use the TI-83 to simulate tossing a coin 999 times. (The maximum number of tosses allowed on the calculator is 999). You will designate "0" as Heads and "1" as Tails.

Press MATH, highlight **PRB**, and select **5:randInt(** and press ENTER. The **randInt(** command requires a minimum value, (which is 0 for this simulation), a maximum value (which is 1), and the number of trials (999). In the **randInt(** command type in **0** ⏹, **1** ⏹, **999.**

```
randInt(0,1,999)
```

Press ENTER. It will take several seconds for the calculator to generate 999 tosses. Notice, in the upper right hand corner a flashing ⏹, indicating that the calculator is working. When the simulation has been completed, a string of **0's** and **1's** will appear on the screen followed by **....,** indicating that there are more numbers in the string that are not shown.

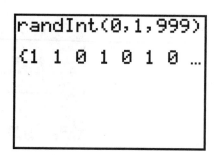

```
randInt(0,1,999)
{1  1  0  1  0  1  0 ...
```

Press **STO** and **2nd** **[L1]** **ENTER** to store the string of numbers in **L1**. Press **2nd** **[LIST]** and highlight **MATH**. Select **sum(** and type in **L1**.

```
randInt(0,1,999)
{1  1  0  1  0  1  0 ...
Ans→L₁
{1  1  0  1  0  1  0 ...
sum(L₁)
                483
```

The sum of **L1** equals the number of Tails in the list. For this particular simulation, the sum is 483. The number of Heads is equal to (999 - no. of Tails). The proportion of Heads is (no. of Heads / 999). How close is this proportion to 50% ?

Section 3.2

▶ Exercise 25 d (pg.123) Birthday Problem

To simulate this problem, press MATH, highlight **PRB**, select **5:randInt(** and enter **1** [,] **365** [,] **24.**

```
randInt(1,365,24
)
{250 364 159 26…
```

Store the data in **L1** by pressing STO and 2nd [L1] ENTER. To see if there are at least two people with the same birthday, you must look for a matching pair of numbers in **L1**. To do this, press 2nd [LIST], highlight **OPS** and select **1:SortA(** and press ENTER. The column you want to sort in ascending order is **L1**, so type 2nd [L1] into the sort command and press ENTER.

```
randInt(1,365,24
)
{292 351 153 60…
Ans→L1
{292 351 153 60…
SortA(L1)
              Done
```

Press STAT and select **1:EDIT**, press ENTER and scroll down through **L1** and check for matching numbers. If you find any matching numbers, that means that at least 2 people in your simulation have the same birthday.

```
L1        L2       L3        1
15    ------   ------
15
23
44
60
64
74
L1(1)=15
```

Notice in this simulation, 15 is listed twice. This represents two people with the same birthday, January 15 (the 15th day of a year).

Section 3.4

▶ Example 3 (pg. 135) Finding the Number of Permutations

You have a baseball team consisting of nine players. To find the number of different batting orders that are possible, you must calculate the number of permutations of **9** objects taken **9** at a time. The formula **nPr** is used with **n = 9** (the number of players on the team) and **r = 9** (the number of players you will be selecting). So, the formula is **9P9**.

Press **9**, **MATH**, highlight **PRB** and select **2:nPr** and **ENTER**.

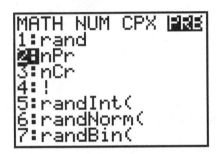

Now press **9** and **ENTER**. The answer, 362880, appears on the screen.

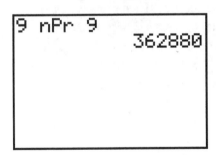

▶ Example 5 (pg. 136) More Permutations

Use the permutation formula with **n = 43** and **r = 3.** Enter **43,** press MATH, highlight **PRB**, select **2:nPr** and press ENTER. Now enter **3** and press ENTER.

```
43 nPr 3
              74046
```

◀

▶ Example 6 (pg. 137) Distinguishable Permutations

To calculate $\dfrac{12!}{6!4!2!}$ you will use the factorial function (**!**). Enter the first value, **12**, press MATH, highlight **PRB** and select **4:!** . Then press ÷ . Open the parentheses by pressing ((. Enter the next value, **6**, press MATH, **PRB**, and select **4:!** . To multiply by 4, press x and enter the next value, **4**. Press MATH, **PRB**, and select **4:!** . To multiply by 2, press x and enter the next value, **2**. Press MATH, **PRB**, and select **4:!** . Close the parentheses)) and press ENTER.

```
12!/(6!*4!*2!)
             13860
■
```

◀

▶ Example 7 (pg. 138) Finding the Number of Combinations

To calculate the number of different combinations of four companies that can be selected from 16 bidding companies you will use the combination formula: **nCr,** where n = the total number of items in the group and r = the number of items being selected. In this example, the formula is **16C4**. Enter the first value, **16,** press MATH, highlight **PRB** and select **3: nCr,** and enter the second value, **4.** Press ENTER and the answer will be displayed on the screen..

```
16 nCr 4
              1820
```

◀

▶ Example 9 (pg. 139) Finding Probabilities

To calculate the probability of being dealt five diamonds from a standard deck of 52 playing cards, you must calculate: $\dfrac{13C5}{52C5}$

To calculate the numerator, enter the first value, **13**, press MATH, highlight **PRB** and select **3:nCr** and enter the next value, **5**. Next press ÷ and enter the first value in the denominator, **52**, press MATH, highlight **PRB** and select **3:nCr**, enter the next value, **5.** Press ENTER and the answer will be displayed on your screen.

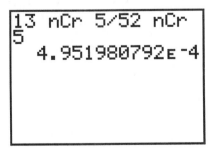

```
13 nCr 5/52 nCr
5
   4.951980792E-4
```

Notice that the answer is written in scientific notation. To convert to standard notation, move the decimal point 4 places to the left. Thus, the answer is .000495.

◀

▶ Exercise 30 (pg. 142) Probability

a. To calculate 200C15, enter the first value, **200**, press MATH, highlight **PRB** and select **3:nCr**, enter the next value, **15** and press ENTER.

b. Calculate 144C15 (follow the steps for part a.).

c. The probability that *no* minorities are selected is equal to the probability that the committee is composed completely of non-minorities. This probability can be calculated with the following formula: $\frac{144C15}{200C15}$.

◀

▶ Technology Lab (pg. 143) Composing Mozart Variations

Exercises 3 - 4: To select one number from 1 to 11, press MATH, highlight
PRB and select **5:randInt(** and enter **1** [,] **11**) and press ENTER. One random
number between 1 and 11 will be displayed on the screen.

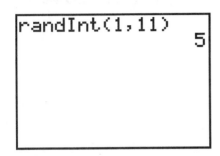

To select 100 integers between 1 and 11, press MATH, highlight **PRB** and select
5:randInt(and enter **1** [,] **11** [,] **100**) and press ENTER. Store the results in
L1 by pressing STO, 2nd [L1], ENTER. Next you can create a histogram and
use it to tally your results. To create the histogram, press 2nd [STAT PLOT],
select **1:Plot 1** and press ENTER. Turn **ON** Plot 1, set **Type** to **Histogram.** Set
Xlist to **L1** and **Freq** to **1.** Press ZOOM and select 9 for **ZoomStat.** Press
WINDOW and set **Xscl** = **1,** then press GRAPH. Use TRACE to scroll through
the bars of the histogram.

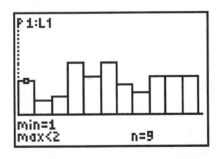

The first bar represents the number **1.** (Notice on the screen: min = 1). Record
the number of items in each bar in a frequency table. In the example you see
here, for X = 1, n = 9. Convert each frequency into a relative frequency and use
the relative frequencies to answer the question in 4b.

Exercise 5: To select 2 random numbers from 1 to 6, press MATH, highlight
PRB and select **5:randInt(** and enter **1** [,] **6** [,] **2**) and press ENTER. Add
the two numbers in the display and subtract **1** from the total.

To select 100 totals, create 2 columns of 100 random numbers. Press MATH, highlight **PRB** and select **5:randInt(** and enter **1** [,] **6** [,] **100**) and press ENTER. Press STO 2nd [L1] and ENTER to store this first string of numbers in **L1**. Next create a second set of 100 numbers using the same procedure. Press STO 2nd [L2] and ENTER to store this second string of numbers in **L2**. Now you can calculate the sum of each row and subtract **1.** Press STAT and select **1:EDIT**. Move the cursor to highlight **L3**.

```
L1        L2        [L3]      3
 1         4        -------
 3         2
 6         4
 2         3
 1         1
 1         2
 3         6
L3 =
```

Press ENTER and the cursor will move to the bottom of the screen. On the bottom line next to **L3 =,** type in 2nd [L1] + 2nd [L2] - 1 and press ENTER. The results should appear in **L3**.

```
L1        L2        L3        3
 1         4        [4]
 3         2         4
 6         4         9
 2         3         4
 1         1         1
 1         2         2
 3         6         8
L3(1)=4
```

Create a histogram of **L3** and use it to tally your results (follow the steps in Exercises 3 - 4 to create the histogram). Use TRACE to find the frequencies for each X-value and record your results in a frequency table. Convert each frequency to a relative frequency and use the relative frequencies to answer question 6b.

Exercise 7: Simulate the tossing of 2 dice 16 times. (Press MATH, highlight **PRB** and select **5:randInt(** and enter **1** [,] **6** [,] **16**). Store the 16 tosses of the first die in **L1** and the 16 tosses of the second die in **L2**. Calculate the sum of each row and store the sums in **L3**.

```
L1      L2      L3      3
2       1       ▓▓▓▓▓▓
3       6       9
5       4       9
6       4       10
4       3       7
5       4       9
5       1       6
L3(1)=3
```

Scroll down to the eighth item in **L3.** If this number is odd, select Option 1 for the eighth bar of the minuet. If it is even, select Option2 for the eighth bar.

```
L1      L2      L3      3
3       6       9
5       4       9
6       4       10
4       3       7
5       4       9
5       1       6
3       5       ▓▓▓▓▓▓
L3(8) =8
```

In this example, the eighth item in **L3** is 8, so the eighth bar of the minuet is Option2. Next, scroll down to the sixteenth item in **L3.** If this number is odd, select Option 1 for the sixteenth bar of the minuet. If it is even, select Option2 for the sixteenth bar.

```
L1      L2      L3      3
2       6       8
3       6       9
1       1       2
2       3       5
4       1       5
6       2       8
5       6       ▓▓▓▓▓▓
L3(16) =11
```

In this example, the sixteenth item in **L3** is 11, so the sixteenth bar of the minuet is Option1.

For each of the other bars in the minuet, subtract **1** from the total. To do this, highlight **L3.** Press ENTER. On the bottom line of the screen, next to **L3** = type in **L3 - 1.** Press ENTER and the new values in **L3** will be displayed on the screen.

```
┌─────┬─────┬─────────┐
│L1   │L2   │L3      3│
├─────┼─────┼─────────┤
│2    │1    │▓▓▓▓▓▓▓  │
│3    │6    │8        │
│5    │4    │8        │
│6    │4    │9        │
│4    │3    │6        │
│5    │4    │8        │
│5    │1    │5        │
├─────┴─────┴─────────┤
│L3(1)=2              │
└─────────────────────┘
```

The first 7 numbers in **L3** represent the first 7 bars of the minuet. Items 9
through 15 in **L3** represent the 9th through 15th bars of the minuet.
Combine these values with bars 8 and 16 that you previously selected and
you have the completed minuet. In this example, the resulting minuet is:
2 8 8 9 6 8 5 2 10 7 8 1 4 4 7 1.

◀

Discrete Probability Distributions

CHAPTER

4

Section 4.1

▶ Example 5 (pg. 156) Mean of a Probability Distribution

Press **STAT** and select **1:EDIT**. Clear **L1** and **L2**. Enter the X-values into **L1** and the P(x) values into **L2**. Press **STAT** and highlight **CALC**. Select **1:1-Var Stats,** press **ENTER** and press 2nd [L1] , 2nd [L2] **ENTER** to see the descriptive statistics. The mean score is 2.94.

```
1-Var Stats
 x̄=2.94
 Σx=2.94
 Σx²=10.26
 Sx=
 σx=1.271377206
↓n=1
```

◀

▶ Example 6 (pg. 157) The Variance and Standard Deviation

Using the data from Example 5 on pg. 156, which you stored in **L1** and **L2**, press STAT highlight **CALC**. Select **1:1-Var Stats,** press ENTER and press 2nd [L1] , 2nd [L2] ENTER. The population standard deviation , σx, is 1.27.

```
1-Var Stats
 x̄=2.94
 Σx=2.94
 Σx²=10.26
 Sx=
 σx=1.271377206
↓n=1
■
```

To find the variance, type in **1.27** and press the x^2 key. The population variance is **1.6129**.

```
 Σx=2.94
 Σx²=10.26
 Sx=
 σx=1.271377206
↓n=1
1.27²
             1.6129
■
```

Section 4.2

▶ Example 5 (pg. 169)   Binomial Probabilities

To find a binomial probability you will use the binomial probability density
function, **binompdf(n,p,x).** For this example, n = 100, p = .58 and x = 65. Press
2nd [DISTR]. Scroll down through the menu to select **0:binompdf(** and press
ENTER . Type in **100** ⌐ **.58** ⌐ **65**) and press ENTER. The answer, **.0299**,
will appear on the screen.

```
binompdf(100,.58
,65)
        .0299216472
■
```

▶ Example 6 (pg. 170) Binomial Probabilities

For this example, n = 4 and p = .41. You want to calculate P(X = 2), P(X ≥ 2) and P(X < 2). Since n is a small number, you can easily display the individual probabilities for each X-value (X = 0,1,2,3,4). To make it easier to view the individual probabilities, change the number of decimal places in the display by pressing MODE and changing **FLOAT** (on line 2 of the display) to **3**.

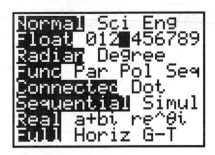

Press ENTER. This will round each of the probabilities to 3 decimal places. Press 2nd [QUIT].

To calculate the individual probabilities, you will use the command: **binompdf(n,p,x).** If you do not enter a value for **x** in this command, the probabilities for all possible X-values will be calculated and displayed on the screen. So, press 2nd [DISTR] and select **0:binompdf(** and type in **4** , **.41**) and press ENTER. The output is **(.121, .337, .351, .163, .028).** Use the right arrow to scroll through the output. These probabilities are P(X = 0), P(X = 1), P(X = 2), P(X = 3), and P(X = 4).

▶ Example 7 (pg. 171) Graphing Binomial Distributions

Construct a probability distribution for a binomial probability model with n = 6 and p = .65. Press STAT, select **1:EDIT** and clear **L1** and **L2**. Enter the values 0 through 6 into **L1**. Press 2nd [QUIT]. To calculate the probabilities for each X-value in **L1**, first change the display mode so that the probabilities displayed will be rounded to 3 decimal places. Press MODE and change from **FLOAT** to **3** and press 2nd [QUIT]. Next press 2nd [DISTR] and select **0:binompdf(** and type in **6** , **.65**) and press ENTER. Store these probabilities in **L2** by pressing STO 2nd [L2] ENTER.

```
binompdf(6,.65)
{.002 .020 .095…
Ans→L₂
{.002 .020 .095…
```

To graph the binomial distribution, press 2nd [STAT PLOT] and press ENTER. Turn **ON** Plot 1, select **Histogram** for **Type**, type in 2nd [L1] for **Xlist** and 2nd [L2] for **Freq.** Press ZOOM and select 9 for **ZoomStat.** Adjust the graph by pressing WINDOW and setting **Xmin = 0, Xmax = 7, Xscl = 1, Ymin = 0** and **Ymax = .35.** Choosing an Xmax=7, leaves some space at the right of the graph in order to complete the histogram. The Ymax value was selected by looking through the values in **L2** and then rounding the largest value UP to a convenient number. Press GRAPH to view the histogram.

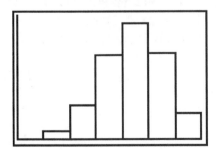

▶ Exercise 7 (pg. 174) Binomial Probabilities

A multiple-choice quiz consists of five questions and each question has four possible answers. If you randomly guess the answer to each question, the probability of getting the correct answer is 1 in 4, so p = .25 and n = 5.

a. To calculate P(x = 3), press 2nd [DISTR] and select **0:binompdf(** and type in **5** [,] **.25** [,] **3** [)] and press ENTER.

b-c. To calculate inequalities, such as P(X ≥ 3) or P(X < 3) , you can use the cumulative probability command: **binomcdf (n,p,x)**. This command accumulates probability starting at X = 0 and ending at a specified X-value. To calculate the probability of guessing at least three answers correctly, that is P(X ≥ 3), press 2nd [DISTR] and select **A:binomcdf(** by scrolling through the options and selecting **A:binomcdf(** or by pressing ALPHA A. (Note: A is the ALPHA function on the MATH key.) Type in **5** [,] **.25** [,] **2** [)] and press ENTER. The result, P(X ≤ 2) = .896.

```
binomcdf(5,.25,2
)
                .896
```

  This result is the accumulated probability that X = 0,1 or 2, since this command, **binomcdf**, accumulates probability starting at X = 0. You want P(X ≥ 3) and this is the *complement* of P(X ≤ 2). So P(X ≥ 3) = 1 - P(X ≤ 2) which is .104. The probability of guessing *less than* 3 answers correctly is P(X ≤ 2) which equals 0.896.

◀

Section 4.3

▶ Example 1 (pg. 178) The Geometric Distribution

Suppose the probability that you will make a sale on any given telephone call is 0.23. To find the probability that your *first* sale on any given day will occur on your fourth or fifth telephone call of the day, press **2ⁿᵈ** **[DISTR]** and select **D:geometpdf(** and type in **.23** **,** **2ⁿᵈ** **{ 4 , 5** **2ⁿᵈ** **})** . This command will display 2 probabilites: P(X = 4) and P(X = 5). The probability that your first sale of the day will occur on your fourth or fifth sales call of the day is .105003 + .080852.

```
geometpdf(.23,{4
,5})
{.10500259 .080…
■
```

▶ Example 2 (pg. 179)   The Poisson Distribution

Use the command **poissonpdf (μ,x)** with $\mu = 3$ and x = 4. Press **2ⁿᵈ** **[DISTR]** and select **B:poissonpdf(** and type in **3** **,** **4)** and press **ENTER**. The answer will appear on the screen.

```
poissonpdf(3,4)
       .1680313557
```

▶ Exercise 9 (pg. 183)   Poisson Distribution

This is a Poisson probability problem with μ = 8.

a.   To find the probability that exactly 4 businesses will fail in any given hour, press 2nd [DISTR] and select **B:poissonpdf(** . Type in **8** ⎵ **4** ⎵ and press ENTER.

b.   Use the cumulative probability command **poissoncdf(.** This command accumulates probability starting at X = 0 and ending at the specified X-value. To calculate the probability that at least 4 businesses will fail in any given hour, that is P(X ≥ 4), you must find P(X ≤ 3) and subtract from 1. Press 2nd [DISTR] , select **C:poissoncdf(** and type in **8** ⎵ **3** ⎵ and press ENTER. The result, P(X ≤ 3), is .0424. This result is the accumulated probability that X = 0,1,2, or 3. P(X ≥ 4) is the *complement* of P(X ≤ 3). So, P(X ≥ 4) is 1 - .0424 or .9576.

```
poissoncdf(8,3)
       .042380112
```

c.   To find the probability that more than 4 businesses will fail in any given hour, find P(X > 4). So, first calculate the *complement*, P(X ≤ 4), and subtract the answer from 1.

```
poissoncdf(8,4)
        .0996324005
█
```

▶ Technology Lab (pg. 185)

Exercise 1: Create a Poisson probability distribution with $\mu = 4$ for $X = 0$ to 10. Press STAT and select **1:EDIT** . Clear **L1** and **L2**. Enter the integers 0,1,2,3,...10 into **L1** and press 2nd [QUIT].

For this example, it is helpful to display the probabilities with 3 decimal places. Press MODE. Move the cursor to the 2nd line and select **3** and press ENTER.

Press 2nd [DISTR] and select **B:poissonpdf(** . Type in **4** , 2nd [L1] and press ENTER .

```
poissonpdf(4,L₁)
{.018 .073 .147…
■
```

A partial display of the probabilities will appear. Press STO and 2nd [L2] ENTER. Press STAT, and select **1:EDIT. L1** and **L2** will be displayed on the screen.

```
L1        L2       L3      1
0.000     .018     ------
1.000     .073
2.000     .147
3.000     .195
4.000     .195
5.000     .156
6.000     .104
L1(1)=0
```

Notice that the X-values in **L1** now have 3 decimal places. This is a result of setting the MODE to **3.** Each value in **L2** is the poisson probability associated with the X-value in **L1**. So, for example, P(X = 2) is .147. Scroll through **L2** and compare the probabilities in **L2** with the heights if the corresponding bars in the frequency histogram on page 185.

Exercises 3 - 5: Use another technology tool to generate the random numbers. The TI-83 cannot generate poisson random data.

Exercise 6: Use the probability distribution that you generated in Exercise 1. Scroll through **L1** to find X = 10. Read the corresponding probability in **L2**.

Exercise 7: Use the probability distribution that you generated in Exercise 1.

a.   Sum the probabilities in **L2** that correspond to X- values of 3, 4 and 5 in **L1**.

b.   Sum the probabilities in **L2** that correspond to X- values of 0,1,2 and 3 in **L1**. This is $P(X \leq 3)$. Subtract this sum from **1** to get $P(X \geq 4)$.

c.   Assuming that the number of arrivals per minute are independent events, raise $P(X \geq 4)$ to the fourth power.

Normal Probability Distributions

Section 5.3

▶ Example 4 (pg. 215) Normal Probabilities

Suppose that cholesterol levels of American men are normally distributed with a mean of 215 and a standard deviation of 25. If you randomly select one American male, calculate the probability that his cholesterol is less than 175, that is P(X < 175).

The TI-83 has two methods for calculating this probability.

Method 1: **Normalcdf**(lowerbound, upperbound, μ, σ) computes probability between a lowerbound and an upperbound. In this example, you are computing the probability from negative infinity to 175. Negative infinity is specified by [(-)] [1] 2ⁿᵈ [EE] [9] [9] (Note: **EE** is found above the comma [,]).

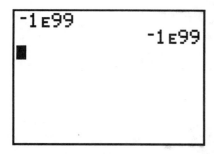

Press 2ⁿᵈ [DISTR] and select **2:normalcdf(** and type in **-1E99** [,] **175** [,] **215** [,] **25** [)] and press ENTER.

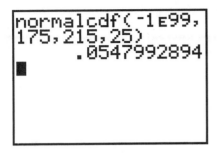

Method 2: This method calculates P(X < 175) and also displays a graph of the probability distribution. You must first set up the WINDOW so that the graph will be displayed properly. You will need to set Xmin equal to (μ - 3 σ) and Xmax equal to (μ + 3 σ). Press WINDOW and set **Xmin** equal to (μ - 3 σ) by entering **215 - 3 * 25.**

```
WINDOW
 Xmin=215-3*25■
 Xmax=290
 Xscl=1
 Ymin=-.005
 Ymax=.02
 Yscl=1
 Xres=1
```

Press ENTER and set **Xmax** equal to (μ + 3 σ) by entering **215 + 3 * 25.** Set **Xscl** equal to σ .

Setting the Y-range is a little more difficult to do. A good "rule - of - thumb" is to set **Ymax** equal to .5 / σ . For this example, type in **.5 / 25** for **Ymax.**

```
WINDOW
 Xmin=140
 Xmax=290
 Xscl=25
 Ymin=-.005
 Ymax=.5/25■
 Yscl=1
 Xres=1
```

Use the blue up arrow to highlight **Ymin.** A good value for **Ymin** is **(-) Ymax / 4** so type in [(-)] **.02 / 4.**

```
WINDOW
 Xmin=140
 Xmax=290
 Xscl=25
 Ymin=-.02/4■
 Ymax=.02
 Yscl=1
 Xres=1
```

Press 2ⁿᵈ [QUIT]. Clear all the previous drawings by pressing 2ⁿᵈ **[DRAW]** and selecting **1:ClrDraw** and pressing ENTER ENTER. Now you can draw the probability distribution. Press 2ⁿᵈ [DISTR]. Highlight **DRAW** and select **1:ShadeNorm(** and type in **-1E99** , **175** , **215** , **25**) and press ENTER. The output displays a normal curve with the appropriate area shaded in and its value computed.

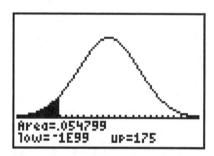

```
Area=.054799
low=-1E99   up=175
```

▶ Example 5 (pg. 216) Finding a Specific Data Value

This is called an inverse normal problem and the command **invNorm(area,** μ, σ) is used. In this type of problem, a percentage of the area under the normal curve is given and you are asked to find the corresponding X-value. In this example, the percentage given is the top 5 %. The TI-83 always calculates probability from negative infinity up to the specified X-value. To find the X-value corresponding to the top 5 %, you must accumulate the bottom 95 % of the area. Press 2nd [DISTR] and select **3:invNorm(** and type in **.95** [,] **75** [,] **6.5**) and press ENTER.

```
invNorm(.95,75,6
.5)
          85.69154857
```

In order to score in the top 5 %, you must earn a score of at least 85.69. Assuming that scores are given as whole numbers, your score must be at least 86.

◀

▶ Exercise 3 (pg. 217) Finding Probabilities

In this exercise, use a normal distribution with $\mu = 69.2$ and $\sigma = 2.9$.

Method 1:To find P(X < 66), press 2ⁿᵈ [DISTR] , select **2:normalcdf(** and type in **-1E99 [,] 66 [,] 69.2 [,] 2.9 [)]** and press ENTER.

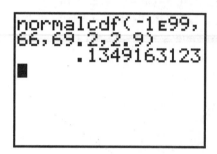

Method 2: To find P(X < 66) and include a graph, you must first set up the Graph Window. Press WINDOW and set **Xmin = 69.2 - 3 * 2.9 and Xmax = 69.2 + 3 * 2.9.** Set **Xscl** = 2.9. Set **Ymax = .5 / 2.9) and Ymin = -.1724/4.** Press 2ⁿᵈ [DRAW] and select **1:ClrDraw** and press ENTER ENTER. Press 2ⁿᵈ [DISTR], highlight **DRAW** and select **1:ShadeNorm(** and type in **-1E99 [,] 66 [,] 69.2 [,] 2.9 [)]** and press ENTER.

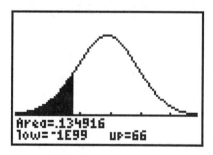

To find p(66<x< 72), press 2ⁿᵈ [DISTR] , select **2:normalcdf(** and type in **66 [,] 72 [,] 69.2 [,] 2.9 [)]** and press ENTER .

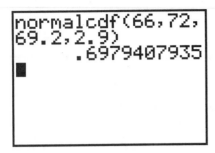

or press 2nd [DRAW] and select **1:ClrDraw** and press ENTER ENTER. Press 2nd [DISTR] , highlight **DRAW** and select **1:ShadeNorm(** and type in **66** , **72** , **69.2** , **2.9**) and press ENTER .

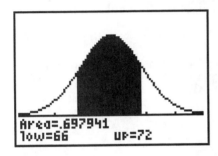

To find P(X > 72), press 2nd [DISTR] , select **2:normalcdf(** and type in **72** , **1E99** , **69.2** , **2.9**) and press ENTER. (Note: In this example, the lowerbound is 72 and the upperbound is positive infinity).

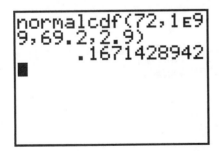

or press 2nd [DRAW] and select **1:ClrDraw** and press ENTER ENTER. Press 2nd [DISTR] , highlight **DRAW** and select **1:ShadeNorm(**and type in **72** , **1E99** , **69.2** , **2.9**) and press ENTER.

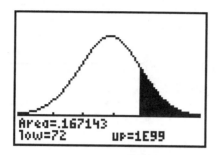

▶ Exercise 9 (pg. 218) Normal Distributions

This is the normal distribution with $\mu = 69.2$ and $\sigma = 2.9$ from Exercise 3 on pg. 217.

a. To calculate P(X > 75) press 2ⁿᵈ [DISTR] , select **2:normalcdf(** and type in
    75 ⃞ , ⃞ **1E99** ⃞ , **69.2** ⃞ , **2.9** ⃞ and press ENTER.

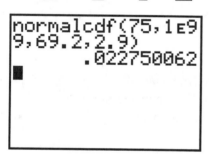

b. To calculate P(X < 72), press 2ⁿᵈ [DISTR] , select **2:normalcdf(** and type
    in **-1E99** ⃞ , **72** ⃞ , **69.2** ⃞ , **2.9** ⃞ and press ENTER.

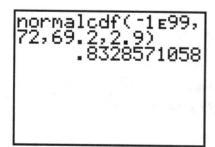

So, if 100 men are selected, the number of men who will have a height less than 72 is **.8329 * 100** which is **83.29** or approximately **83** men.

c. To find the 90th percentile, press 2nd [DISTR] and select **3:invNorm(** and type in **.90** , **69.2** , **2.9**) and press ENTER.

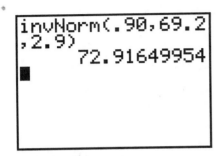

d. To find the first quartile, press 2nd [DISTR] and select **3:invNorm(** and type in **.25** , **69.2** , **2.9**) and press ENTER .

```
invNorm(.25,69.2
,2.9)
        67.24397973
■
```

◀

Section 5.4

▶ Example 4 (pg. 227) Probabilities for Sampling Distributions

In this example, a sample of n = 50 adults ages 18 to 24 is selected and the length of time each person spends reading is recorded. The mean of the population from which the sample was selected is $\mu = 9$ minutes and the standard deviation is $\sigma = 1.5$ minutes. Since n > 30, you can conclude that the sampling distribution of the sample mean is approximately normal with $u_{\bar{x}} = 9$ and $\sigma_{\bar{x}} = 1.5/\sqrt{50}$. To calculate P($8.7 < \bar{x} < 9.5$), press 2nd [DISTR] , select **2:normalcdf(** and type in **8.7** ⬚ **9.5** ⬚ **9** ⬚ **1.5/**$\sqrt{50}$ ⬚ and press ENTER.

```
normalcdf(8.7,9.
5,9,1.5/√(50)
        .9121393013
```

▶ Example 6 (pg. 229) Finding Probabilities for x and $\bar{x}$

The population is normally distributed with $\mu = 2870$ and $\sigma = 900$.

1. To calculate P(X <2500), press 2nd [DISTR] , select **2:normalcdf(** and type
in **-1E99** ⬚**,** **2500** ⬚**,** **2870** ⬚**,** **900** ⬚**)** and press ENTER.

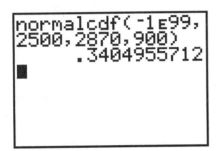

```
normalcdf(-1E99,
2500,2870,900)
       .3404955712
■
```

2. To calculate P($\bar{x}$ <2500), press 2nd [DISTR] , select **2:normalcdf(** and type
in **-1E99** ⬚**,** **2500** ⬚**,** **2870** ⬚**,** **900/$\sqrt{25}$** ⬚**)** and press ENTER.

```
normalcdf(-1E99,
2500,2870,900/√(
25))
      .0199126205
■
```

◀

▶ Exercise 21 (pg. 232) Make a Decision

To calculate P($\bar{x}$ <127.9), press 2nd [DISTR] , select **2:normalcdf(** and type in
-1E99 [,]   **127.9** [,]   **128** [,]   **0.20/**$\sqrt{40}$ [ )] and press ENTER.

```
normalcdf(-1E99,
127.9,128,.20/√(
40))
    7.827669991E-4
```

Notice that the answer is displayed in scientific notation: 7.827E-4. Convert
this to standard notation, .0007827, by moving the decimal point 4 places to the
left. This probability is extremely small; therefore, the event ($\bar{x}$ <127.9) is
highly unlikely if the mean is actually 128. So, something has gone wrong with
the machine and the actual mean must have shifted to a value less than 128.

◀

▶ Technology Lab (pg. 234) Age Distribution in the U. S.

Exercise 1: Press **STAT** and select **1:EDIT**. Clear **L1, L2** and **L3**.
Enter the age distribution into **L1** and **L2**. Put class midpoints in **L1** and relative
frequencies (converted to decimals) into **L2**. To find the population mean, μ ,
and the population standard deviation, σ , press **STAT**, highlight **CALC**, select
1:1-Var Stats, press **ENTER** and press 2nd [L1] , 2nd [L2] **ENTER**. The
mean and the standard deviation will be displayed. (Note: Use σ x because the
age data represents the entire population distribution of ages, not a sample.)

Exercise 2: Enter the thirty-six sample means into **L3**. To find the mean
and standard deviation of these sample means, press **STAT**, highlight
CALC, select **1:1-Var Stats**. Press **ENTER** and press 2nd [L3] **ENTER**.
The mean and the standard deviation will be displayed. (Note: use Sx
because the 36 sample means are a sample of 36 means, not the entire
population of all possible means of size n = 40).

Exercise 3: Construct a histogram of the age distribution. Press 2nd
[STAT PLOT] and select **1: Plot 1**. Turn **ON** Plot 1, select **Histogram**
for **Type,** set **Xlist** to **L1** and set **Freq** to **L2**. Press **ZOOM** and **9** for
ZoomStat. Adjust the WINDOW by pressing **WINDOW** and setting
Xmax = 102.5, Xscl = 5, Ymin = -.02 and Ymax = .09.

Exercise 4: The TI-83 will draw a frequency histogram, not a *relative*
frequency histogram. (The shape of the data can be determined from either
type of histogram). Press 2nd **[STAT PLOT]** and select **1: Plot 1. Plot 1**
has already been turned **ON** and **Histogram** has been selected for **Type**.
Set **Xlist** to **L3**. On the **Freq** line, press **CLEAR** and **ALPHA** to return to
the standard rectangular cursor and enter **1**. Press **ZOOM** and **9** for
ZoomStat. To adjust the histogram so that it has nine classes, press
WINDOW. Approximate the class width using (48 - 28)/9. This value is
approximately 2, so set **Xscl = 2** and press **GRAPH**.

Exercise 5: See the output from Exercise 1 for the population standard
deviation.

Exercise 6: See the output from Exercise 2 for the standard deviation of
the sample means.

CHAPTER

Confidence Intervals

6

Section 6.1

▶ Example 4 (pg. 256)   Constructing a Confidence Interval

Enter the data from Example 1 on pg. 252 into **L1**. In this example, n > 30, so
the sample standard deviation, Sx, is a good approximation of σ, the population
standard deviation. To find Sx, press **STAT**, highlight **CALC**, select **1:1-Var
Stats** and press **ENTER** **ENTER**.

```
1-Var Stats
 x̄=12.42592593
 Σx=671
 Σx²=9671
 Sx=5.015454801
 σx=4.968798394
↓n=54
```

The sample standard deviation is Sx = 5.02. Using this as an estimate of σ, you
can construct a Z-Interval, a confidence interval for μ, the population mean.
Press **STAT**, highlight **TESTS** and select **7:Zinterval.**

```
EDIT CALC TESTS
1:Z-Test…
2:T-Test…
3:2-SampZTest…
4:2-SampTTest…
5:1-PropZTest…
6:2-PropZTest…
7↓ZInterval…
```

On the first line of the display, you can select **Data** or **Stats.** For this example,
select **Data** because you want to use the actual data which is in **L1**. Press
ENTER . Move to the next line and enter 5.02, your estimate of σ. On the next

line, enter **L1** for **LIST**. For **Freq**, enter **1**. For **C-Level** , enter **.99** for a 99% confidence interval. Move the cursor to **Calculate.**

```
ZInterval
 Inpt:Data Stats
 σ:5.02
 List:L₁
 Freq:1
 C-Level:.99
 Calculate
```

Press **ENTER** .

```
ZInterval
 (10.666,14.186)
 x̄=12.42592593
 Sx=5.015454801
 n=54
```

A 99% confidence interval estimate of μ , the population mean is (10.666, 14.186). The output display includes the sample mean (12.43), the sample deviation (5.015), and the sample size (54).

▶ Example 5 (pg. 257) Confidence Interval for μ (σ known)

In this example, $\bar{x} = 22.9$ years, n = 20, and a value for σ from previous studies is given as $\sigma = 1.5$ years. Construct a 90% confidence interval for μ, the mean age of all students currently enrolled at the college.

Press **STAT**, highlight **TESTS** and select **7:ZInterval**. In this example, you do not have the actual data. What you have are the summary statistics of the data, so select **Stats** and press **ENTER**. Enter the value for σ : **1.5**; enter the value for $\bar{x}$: **22.9**; enter the value for **n: 20**; and enter **.90** for **C-level**. Highlight **Calculate**.

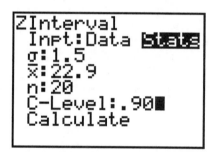

Press **ENTER** .

```
ZInterval
 (22.348,23.452)
 x̄=22.9
 n=20



■
```

Using a 90% confidence interval, you estimate that the average age of all students is between 22.348 and 23.452.

▶ Exercise 31 (pg. 261) Applying the Concept

Enter the data into **L1**. Press **STAT**, highlight **TESTS** and select **7:ZInterval**.
Since you have entered the actual data points into **L1**, select **Data** for **Inpt** and
press **ENTER**. Enter the value for σ, **1.5**. Set **LIST = L1, Freq = 1** and **C-level = .90**. Highlight **Calculate.**

```
ZInterval
 Inpt:DATA Stats
 σ:1.5
 List:L₁
 Freq:1
 C-Level:.9
 Calculate
```

· Now press **ENTER**.

```
ZInterval
 (8.4296,9.7037)
 x̄=9.066666667
 Sx=1.579632266
 n=15

■
```

Repeat the process and set **C-level = .99**.

```
ZInterval
 (8.0691,10.064)
 x̄=9.066666667
 Sx=1.579632266
 n=15

■
```

◀

▶ Exercise 43 (pg. 263) Using Technology

Enter the data into L1. Since no estimate of σ is given, use Sx, the sample standard deviation. This is a good approximation of σ since n > 30. Press **STAT**, highlight **CALC**, select **1:1-Var Stats** and press **ENTER** **ENTER** . The sample standard deviation, Sx, is 11.19.

Press **STAT**, highlight **TESTS** and select **7:ZInterval**. For **Inpt**, select **Data**. For σ , enter **11.19**. Set **List** to **L1**, **Freq** to **1** and **C-level** to **.95**. Highlight **Calculate** and press **ENTER** .

```
ZInterval
 (303.5,311.25)
 x̄=307.375
 Sx=11.18971302
 n=32
```

Section 6.2

▶ Example 3 (pg. 269) Constructing a Confidence Interval

In this example, n = 20, $\bar{x}$ = 6.93 and s = 0.42 and the underlying population is
assumed to be normal. Find a 99% confidence interval for μ. First, notice that
σ is unknown. The sample standard deviation, s, is not a good approximation of
σ when n < 30. To construct the confidence interval for μ, the correct
procedure under these circumstances (n < 30, σ is unknown and the population
is assumed to be normally distributed) is to use a T-Interval.

Press **STAT**, highlight **TESTS**, scroll through the options and select
8:TInterval and press **ENTER** . Select **Stats** for **Inpt** and press **ENTER**. Fill
in $\bar{x}$, Sx, and n with the sample statistics. Set **C-level** to **.99**. Highlight
Calculate.

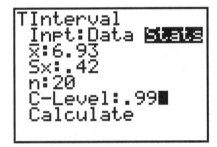

```
TInterval
 Inpt:Data Stats
 x̄:6.93
 Sx:.42
 n:20
 C-Level:.99▮
 Calculate
```

Press **ENTER** .

```
TInterval
 (6.6613,7.1987)
 x̄=6.93
 Sx=.42
 n=20
```

▶ Exercise 21 (pg. 273) Deciding on a Distribution

In this example, n = 70, $\bar{x}$ = 1.25 and s = 0.01. Notice that σ is unknown. You can use s as a good approximation of σ in this case because n > 30. To calculate a 95 % confidence interval for μ , press **STAT**, highlight **TESTS** and select **7:ZInterval**. Fill in the screen with the appropriate information.

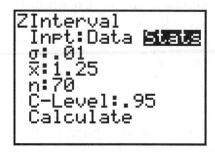

Calculate the confidence interval.

```
ZInterval
 (1.2477,1.2523)
 x̄=1.25
 n=70

■
```

▶ Exercise 23 (pg. 273) Deciding on a Distribution

Enter the data into **L1**. In this case, with n = 25 and σ unknown, assuming that the population is normally distributed, you should use a Tinterval. Press **STAT**, highlight **TESTS** and select **8:TInterval**. Fill in the appropriate information.

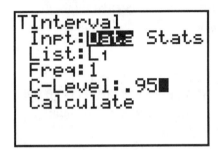

```
TInterval
 Inpt:DATA Stats
 List:L1
 Freq:1
 C-Level:.95■
 Calculate
```

Calculate the confidence interval.

```
TInterval
 (22.762,25.238)
 x̄=24
 Sx=3
 n=25

■
```

Section 6.3

▶ Example 2 (pg. 277)   Constructing a Confidence Interval for p

In this example, 883 American adults are surveyed and 380 say that their favorite sport is football. Construct a 95 % confidence interval for p, the proportion of all Americans who say that their favorite sport is football.

Press **STAT**, highlight **TESTS**, scroll through the options and select **A:1-PropZInt**. The value for X is the number of American adults in the group of 883 who said that football was their favorite sport, so **X = 380**. The number who were surveyed is n, so **n = 883**. Enter **.95** for **C-level**.

```
1-PropZInt
 x:380
 n:883
 C-Level:.95
 Calculate
```

Highlight **Calculate** and press **ENTER** .

```
1-PropZInt
 (.39769,.46301)
 p=.4303510759
 n=883
```

In the output display the confidence interval for p is (.39769, .46301). The sample proportion, $\hat{p}$, is .4304 and the number surveyed is 883.

▶ Example 3 (pg. 278) Confidence Interval for p

From the graph, the sample proportion, $\hat{p}$, is 0.45 and n is 935. Construct a 99% confidence interval for the proportion of adults who think that airplanes are the safest mode of transportation. In order to construct this interval using the TI-83, you must have a value for X, the number of people in the study who said that airplanes were the safest mode of transportation. Multiply 0.45 by 935 to get this value. If this value is not a whole number, round to the nearest whole number. For this example, X is 421.

Press **STAT**, highlight **TESTS** and select **A:1-PropZInt** by scrolling through the options or by simply pressing **ALPHA** **A**. Enter the appropriate information from the sample.

```
1-PropZInt
 x:421
 n:935
 C-Level:.99
 Calculate
```

Highlight **Calculate** and press **ENTER**.

```
1-PropZInt
 (.40836,.49218)
 p=.4502673797
 n=935
```

▶ Exercise 21 (pg. 282) Confidence Interval for p

a. To construct a 99% confidence interval for the proportion of men who favor
    irradiation of red meat, multiply .61 by 500 to obtain a value for X, the
    number of men in the survey who favor irradiation of red meat. Press
    STAT, highlight **TESTS** and select **A:1-PropZInt**. Fill in the appropriate
    values.

```
1-PropZInt
 x:305
 n:500
 C-Level:.99
 Calculate
```

Highlight **Calculate** and press **ENTER** .

```
1-PropZInt
 (.55381,.66619)
 p̂=.61
 n=500

■
```

b. To construct a 99 % confidence interval for the proportion of women who
    favor irradiation of red meat, multiply .44 by 500 to obtain a value for X, the
    number of women in the survey who favor irradiation of red meat. Press
    STAT, highlight **TESTS** and select **A:1-PropZInt**. Fill in the appropriate
    values. Highlight **Calculate** and press **ENTER** .

```
1-PropZInt
 (.38282,.49718)
 p̂=.44
 n=500

■
```

Notice that the two confidence intervals do not overlap. It is therefore unlikely that the proportion of males in the population who favor irradiation of red meat is the same as the proportion of females in the population who favor the irradiation of red meat.

◀

Technology Lab (pg. 283) "Most Admired" Polls

1. Use the survey information to construct a 95 % confidence interval for the proportion of people who would have chosen President Clinton as their most admired man. Press **STAT**, highlight **TESTS** and select **A:1-PropInt**. Set **X = 141** and **n = 1005**. Set **C-level** to **.95**, highlight **Calculate** and press **ENTER**.

3. To construct a 95 % confidence interval for the proportion of people who would have chosen Oprah Winfrey as their most admired female, you must calculate X, the number of people in the sample who chose Oprah. Multiply .06 by 1005 and round your answer to the nearest whole number. Press **STAT**, highlight **TESTS**, select **A:1-PropZInt** and fill in the appropriate information. Press **Calculate** and press **ENTER**.

4. To do one simulation, press **MATH**, highlight **PRB**, select **7:randBin(** and type in **1005** **,** **.07** **)** . The output is the number of successes in a survey of n = 1005 people. In this case, a "success" is choosing Oprah Winfrey as your most admired female. (Note: It takes approximately one minute for the TI-83 to do the calculation).

To run the simulation ten times use **7:randBin(1005,.07,10)**. (These simulations will take approximately 8 minutes). The output is a list of the number of successes in each of the 10 surveys. Calculate $\hat{p}$ for each of the surveys. $\hat{p}$ is equal to (number of successes)/1005.

◀

Hypothesis Testing with One Sample

CHAPTER

7

Section 7.2

▶ Example 8 (pg. 322) Hypothesis Testing Using P-values

The hypothesis test, $H_o : \mu \geq 30$ vs. $H_a : \mu < 30$, is a left-tailed test. The sample statistics are $\bar{x} = 28.5$, s = 3.5 and n = 36. The sample size is greater than 30, so the **Z-Test** is the appropriate test. To run the test, press **STAT**, highlight **TESTS** and select **1:Z-Test**. Since you are using sample statistics for the analysis, select **Stats** for **Inpt** and press **ENTER**. For μ_0 enter 30, the value for μ in the null hypothesis. For σ enter **s**, the sample standard deviation. Enter **28.5** for $\bar{x}$ and **36** for **n**. On the next line, choose the appropriate alternative hypothesis and press **ENTER**. For this example, it is $< \mu_0$, a left-tailed test.

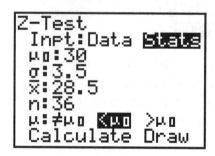

There are two choices for the output of this test. The first choice is **Calculate**. The output displays the alternative hypothesis, the calculated z-value, the P-value, $\bar{x}$ and n.

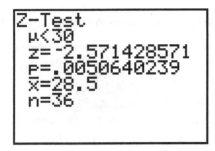

Since p = .005, which is less than α , the correct conclusion is to **Reject** H_o.

To view the second output option, press **STAT**, highlight **TESTS**, and select
1:Z-Test. All the necessary information for this example is still stored in the
calculator. Scroll down to the bottom line and select **DRAW**. A normal curve is
displayed with the left-tail area of .0051 shaded. This shaded area is the area to
the left of the calculated Z-value. The Z-value and the P-value are also
displayed.

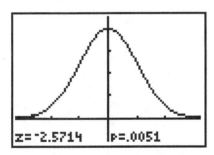

▶ Example 9 (pg. 323)   Hypothesis Testing Using P-values

This test is a two-tailed test for $H_o : \mu = 143260$ vs. $H_a : \mu \neq 143260$. The sample statistics are $\bar{x} = 135000$, s = 30000 and n = 30. Press **STAT**, highlight **TESTS** and select **1:Z-Test**. Choose **Stats** for **Inpt** and press **ENTER**. For μ_0 enter 143260, the value for μ in the null hypothesis. For σ, enter **s**, the sample standard deviation. Enter **135000** for $\bar{x}$ and **30** for **n**. On the next line, choose the appropriate alternative hypothesis and press **ENTER**. For this example, it is $\neq \mu_0$, a two-tailed test.

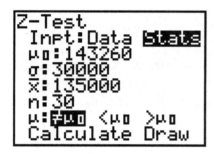

Highlight **Calculate** and press **ENTER** .

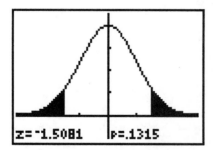

Or, highlight **Draw** and press **ENTER**.

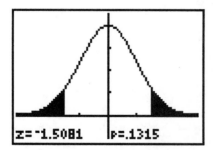

Notice the P-value is equal to .1315. In this example, α is .05. Since the P-value is greater than α, the correct conclusion is to **Fail to Reject** H_o.

▸ Exercise 25 (pg. 326) Testing Claims

This is a hypothesis test for $H_o : \mu \le 28$ vs. $H_a : \mu > 28$. Since n > 30, the appropriate test is the Z-Test. This procedure requires a value for σ. In cases with n > 30, the sample standard deviation, s, can be used to approximate σ. Begin the analysis by entering the 36 data points into **L1**. Press **STAT**, highlight **CALC** and choose **1:1-Var Stats** and press **ENTER ENTER**. The sample statistics will be displayed on the screen. The value you need for the hypothesis test is Sx, the sample standard deviation, which is 22.13.

To run the hypothesis test, press **STAT**, highlight **TESTS** and select **1:Z-Test**. Since you have the actual data points for the analysis, select **Data** for **Inpt** and press **ENTER**. Enter **28** for μ_0 and enter **22.13** for σ. The data is stored in **L1** and **Freq** is **1**. The alternate hypothesis is a right-tailed test so select $> \mu_0$ and press **ENTER**.

```
Z-Test
 Inpt:Data Stats
 μo:28
 σ:22.13
 List:L1
 Freq:1
 μ:≠μo <μo >μo
 Calculate Draw
```

To run the test, select **Calculate** and press **ENTER**.

```
Z-Test
 μ>28
 z=1.317969574
 P=.0937569803
 x̄=32.86111111
 Sx=22.12839517
 n=36
```

Or, select **Draw** and press **ENTER**.

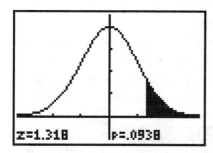

Since the P-value of .0938 is greater than α, the correct conclusion is to **Fail to Reject** H_o.

(Note: the TI-83 does not display critical values. Use the Z-table to find these values).

▶ Exercise 29 (pg. 327) Testing Claims Using P-values

Test the hypotheses: $H_o : \mu \le 260$ vs. $H_a : \mu > 260$.   The sample statistics are $\bar{x} = 265$, s = 55 and n = 85. Press **STAT**, highlight **TESTS** and select **1:Z-Test**. For **Inpt**, choose **Stats** and press **ENTER**. Fill the input screen with the appropriate information. Choose $> \mu_0$ for the alternative hypothesis and press **ENTER**. Highlight **Calculate** and press **ENTER**.

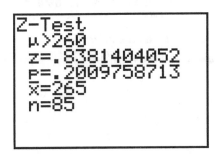

Or, highlight **Draw** and press **ENTER**.

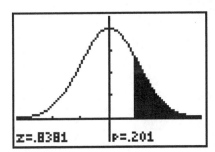

The test statistic is Z = 0.838 and the P-value is .201. Since the P-value is greater than α, the correct conclusion is to **Fail to Reject** H_o.

▶ Exercise 33 (pg. 328) Testing Claims Using P-values

Enter the data into **L1**. Notice that there is no one-tailed alternative hypothesis specified in this exercise. The test, therefore, is run as a two-tailed test and the hypotheses are: $H_o: \mu = 15$ vs. $H_a: \mu \neq 15$. You will need to find the sample standard deviation because the Z-test requires a value for σ. (Recall: Press **STAT**, highlight **CALC** and select **1:1-Var Stats.**) To perform the test, press **STAT**, highlight **TESTS** and select **1:Z-Test**. Since you have the actual data points for the analysis, select **Data** for **Inpt** and press **ENTER**. Fill in the input screen with the appropriate information and select ≠ μ_0, as the alternative hypothesis and press **ENTER**. Highlight **Calculate** and press **ENTER**

```
Z-Test
 μ≠15
 z=-.2183954511
 p=.8271210783
 x̄=14.834375
 Sx=4.287612267
 n=32
■
```

Or, highlight **Draw** and press **ENTER**.

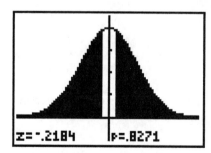

```
z=-.2184        p=.8271
```

Since the P-value is greater than α, the correct conclusion is to **Fail to Reject** H_o.

◀

Section 7.3

▶ Example 6 (pg. 335)   Using P-values with a T-Test

This test is a two-tailed test of $H_o : \mu = 124$ vs. $H_a : \mu \neq 124$. The sample
statistics are $\bar{x} = 135$, s = 20 and n = 11. Since n < 30, the test is a T-Test if you
assume that the underlying population is approximately normally distributed.
Press **STAT**, highlight **TESTS** and select **2:T-Test**. Choose **Stats** for **Inpt** and
press **ENTER**. Fill in the following information: μ_0 = **124,** $\bar{x}$ = **135, Sx = 20**
and **n = 11**. Choose the two-tailed alternative hypothesis, $\neq \mu_0$, and press
ENTER. Highlight **Calculate** and press **ENTER**

```
T-Test
 µ≠124
 t=1.824143635
 P=.0981105275
 x̄=135
 Sx=20
 n=11
■
```

Or, highlight **Draw** and press **ENTER**.

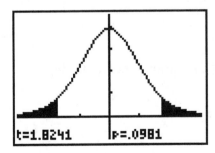

```
t=1.8241        P=.0981
```

Since the P-value is greater than α , the correct conclusion is to **Fail to Reject**
H_o .

▶ Exercise 23 (pg. 338) Testing Claims Using P-values

The correct hypothesis test is $H_o : \mu = 3$ vs. $H_a : \mu < 3$. Enter the data into L1.
Since n < 30, the appropriate test is the T-Test, if you assume that the underlying
population is approximately normal. Press **STAT**, highlight **TESTS** and select
2:T-Test. Select **Data** for **Inpt** and press **ENTER**. Fill in the screen with the
necessary information, choose **Calculate** and press **ENTER**.

```
T-Test
 μ<3
 t=-1.161269599
 P=.1299571833
 x̄=2.785
 Sx=.8279810575
 n=20
```

Or, choose **Draw** and press **ENTER**.

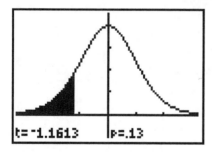

```
t=-1.1613     P=.13
```

Since the P-value is greater than α, the correct conclusion is to **Fail to Reject**
H_o.

▶ Exercise 29 (pg. 339) Deciding on a Distribution

This test is a left-tailed test of $H_o : \mu \geq 21$ vs. $H_a : \mu < 21$. The sample statistics are $\bar{x} = 19$, s = 4 and n = 5. Since n < 30 and gas mileage is normally distributed, the appropriate test is the **T-Test**. Press **STAT**, highlight **TESTS** and select **2:T-Test**. Fill in the screen with the necessary information and choose **Calculate** and press **ENTER**.

```
T-Test
 μ<21
 t=-1.118033989
 p=.1630821177
 x=19
 Sx=4
 n=5
```

Or, choose **Draw** and press **ENTER**.

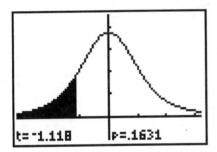

t=-1.118   p=.1631

Since the P-value is greater than α, the correct conclusion is to **Fail to Reject** H_o.

◀

▶ Exercise 31 (pg. 339) Deciding on a Distribution

This test is a left-tailed test of $H_o : \mu \geq 21$ vs. $H_a : \mu < 21$. The sample statistics are $\bar{x} = 19$ and $n = 5$. Since the population standard deviation is known, $\sigma = 5$, the appropriate test is the **Z-Test**. Press **STAT**, highlight **TESTS** and select **1:Z-Test**. Fill in the screen with the necessary information and choose **Calculate** and press **ENTER**.

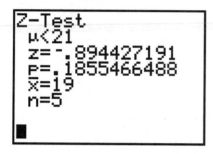

Or, choose **Draw** and press **ENTER**.

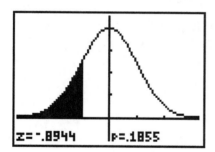

Since the P-value is greater than α, the correct conclusion is to **Fail to Reject** H_o.

Section 7.4

▶ Example 1 (pg. 341)   Hypothesis Test for a Proportion

This hypothesis test is a left-tailed test of: $H_o : p \geq .20$ vs. $H_a : p < .20$. The sample statistics are $\hat{p} = .15$ and n = 100. Press **STAT**, highlight **TESTS** and select **5:1-PropZTest**. This test requires a value for p_0, which is the value for p in the null hypothesis. Enter **.20** for p_0. Next, a value for X is required. X is the number of "successes" in the sample. In this example, a success is " having an allergy". Since 15% of the individuals in the sample say that they have an allergy, **X** is equal to .15 times 100 or **15**. Next, enter the value for n. Select $<$ p_0 for the alternative hypothesis and press **ENTER**.

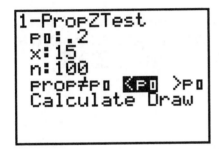

Highlight **Calculate** and press **ENTER**.

```
1-PropZTest
 prop<.2
 z=-1.25
 p=.105649839
 p̂=.15
 n=100
```

Or, highlight **Draw** and press **ENTER**.

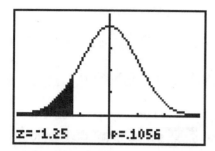

Since the P-value is greater than α, the correct conclusion is to **Fail to Reject** H_o.

◀

> ▶ Example 3 (pg. 343)   Hypothesis Test for a Proportion

The hypothesis test is: $H_o : p \le .55$ vs. $H_a : p > .55$. The sample statistics are $n = 425$ and $X = 255$. Press **STAT**, highlight **TESTS** and select **5:1-PropZTest**. Enter the necessary information. Highlight **Calculate** and press **ENTER**.

```
1-PropZTest
 Prop>.55
 z=2.071938535
 P=.0191355209
 p̂=.6
 n=425
■
```

Or, highlight **Draw** and press **ENTER**.

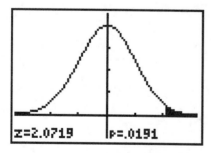

Since the P-value is less than α, the correct conclusion is to **Reject** H_o.

◀

▶ Exercise 9 (pg. 344) Testing Claims

Use a right-tailed hypothesis test to test the hypotheses: $H_o : p \le .30$ vs.
$H_a : p > .30$. The sample statistics are n = 1050 and $\hat{p}$ = .32. Multiply n times
$\hat{p}$ to find X, the number of consumers in the sample who have stopped
purchasing a product because it pollutes the environment. Press **STAT**,
highlight **TESTS** and select **5:1-PropZTest**. Enter the necessary information.
Highlight **Calculate** and press **ENTER**.

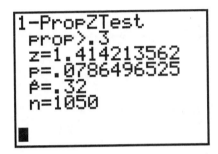

Or, highlight **Draw** and press **ENTER**.

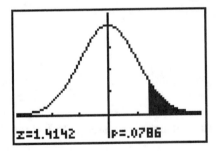

Since the P-value is .greater than α , the correct conclusion is to **Fail to Reject**
H_o .

▶ Exercise 11 (pg. 345) Testing Claims

Use a two-tailed hypothesis test to test: $H_o : p = .60$ vs. $H_a : p \neq .60$. The sample statistics are n = 1762 and X = 1004. Press **STAT**, highlight **TESTS** and select **5:1-PropZTest**. Enter the necessary information. Highlight **Calculate** and press **ENTER**.

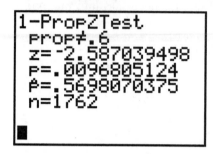

Or, highlight **Draw** and press **ENTER**.

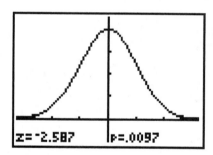

Since the P-value is .0097 and is less than α, the correct conclusion is to **Reject** H_o.

▶ Technology Lab (pg. 346) The Case of the Vanishing Women

Exercise 1: Use the TI-83 to run the hypothesis test and compare your results to the MINITAB results shown in the display at the bottom of pg. 346. Use a two-tailed test to test the hypotheses: $H_o: p = 0.53$ vs. $H_a: p \neq 0.53$. The sample statistics are X = 102 women and n = 350 people selected from the Boston City Directory. Press **STAT**, highlight **TESTS** and select **5:1-PropZTest**. Fill in the appropriate information highlight **Calculate** and press **ENTER** or, highlight **Draw** and press **ENTER**.

Exercise 4: In the first stage of the jury selection process, 350 people are selected and 102 of them are women. So, at this stage, the proportion of women is 102 out of 350, or 0.2914. From this population, 100 people are selected at random and only nine are women. Test the claim that the proportion of women in the population is 0.2914. Use a two-tailed test to test the hypotheses: $H_o: p = 0.2914$ vs. $H_a: p \neq 0.2914$. The sample statistics are X = 9 women and n = 100 people. Press **STAT**, highlight **TESTS** and select **5:1-PropZTest**. Fill in the appropriate information, Highlight **Calculate** and press **ENTER** or highlight **Draw** and press **ENTER**.

◀

Hypothesis Testing with Two Samples

CHAPTER

8

Section 8.1

▶ Example 2 (pg. 372) Two-Sample Z-Test

Test the claim that the average daily cost for meals and lodging when vacationing in Texas is less than the average costs when vacationing in Washington State. Designate Texas as population 1 and Washington as population 2. The appropriate hypothesis test is a left-tailed test of $H_o: \mu_1 \geq \mu_2$ vs $H_a: \mu_1 < \mu_2$. The sample statistics are displayed in the table at the top of pg. 372. Each sample size is greater than 30, so the correct test is a Two-sample Z-Test.

Press **STAT**, highlight **TESTS** and select **3:2-SampZTest**. Since you are using the sample statistics for your analysis, select **Stats** for **Inpt** and press **ENTER**. The first item to enter is a value for σ_1. The sample standard deviation, s_1, can be used as an approximation of σ_1. Enter 15 for σ_1. Enter 28 for σ_2. Next, enter the mean and sample size for group 1: $\bar{x}_1 = 184$ and $n_1 = 50$. Continue by entering the mean and sample size for group 2: $\bar{x}_2 = 195$ and $n_2 = 35$.

```
2-SampZTest
 Inpt:Data Stats
 σ1:15
 σ2:28
 x̄1:184
 n1:50
 x̄2:195
↓n2:35
```

Use the down arrow to display the next line. This line displays the three possible alternative hypotheses for testing: $\mu_1 \neq \mu_2$, $\mu_1 > \mu_2$, or $\mu_1 < \mu_2$. For this example, select $< \mu_2$ and press **ENTER**. Scroll down to the next line and select **Calculate** or **Draw**.

The output for **Calculate** is displayed on two pages. (Notice that the only piece of information on the second page is n_2.)

```
2-SampZTest
  μ1<μ2
  z=-2.12088219
  p=.0169657982
  x̄1=184
  x̄2=195
↓n1=50
```

The output for **Draw** contains a graph of the normal curve with the area associated with the test statistic shaded.

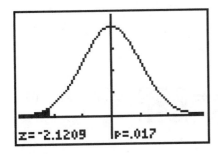

Both outputs display the P-value, which is .017. Since the P-value is greater than α, the correct decision is to **Fail to Reject** H_o.

◄

▶ Exercise 7 (pg. 374) Testing the Difference between Two Means

The appropriate hypothesis test is a two-tailed test of $H_o: \mu_1 = \mu_2$ vs $H_a: \mu_1 \neq \mu_2$. Designate Tire Type A as population 1 and Tire Type B as population 2. Press **STAT**, highlight **TESTS** and select **3:2-SampZTest**. For **Inpt,** select **Stats** and press **ENTER**. Enter the sample standard deviations as approximations of σ_1 and σ_2. Next, enter the sample statistics for each group.

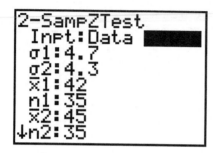

Select $\neq \mu_2$ as the alternative hypothesis and press **ENTER**. Scroll down to the next line and select **Calculate** or **Draw** and press **ENTER**.

The output for **Calculate** is:

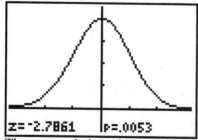

The output for **Draw** is:

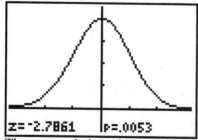

The test statistic, z, is –2.786 and the p-value is .0053. Since the P-value is less than α, the correct decision is to **Reject** H_o. ◀

▶ Exercise 13 (pg. 375) Testing the Difference between 2 Means

Enter the data from 1981 into **L1** and enter the data from the more recent study into **L2**. The hypothesis test is: H_o: $\mu_1 \leq \mu_2$ vs H_a: $\mu_1 > \mu_2$. Since both n_1 and n_2 equal 30, the Two-Sample Z-Test can be used. In order to use this test, values for σ_1 and σ_2 are required. The sample standard deviations, s_1 and s_2 can be used as approximations for σ_1 and σ_2 when both n_1 and n_2 are greater than or equal to 30. To find the standard deviations, press **STAT**, highlight **CALC** and select **1:1-Var Stats** and press **ENTER** **ENTER**. The value for Sx is the sample standard deviation for the 1981 data. Next, press **STAT**, highlight **CALC** and select **1:1-Var Stats**, press **ENTER**, press **2ⁿᵈ [L2]** **ENTER**. The value for Sx is the sample standard deviation for the more recent data.

To run the hypothesis, press **STAT**, highlight **TESTS** and select **3:2-SampZTest**. For **Inpt** select **Data** and press **ENTER**. Enter the sample standard deviations as approximations of σ_1 and σ_2. For **List1**, press **2ⁿᵈ [L1]** and for **List2** press **2ⁿᵈ [L2]** .

```
2-SampZTest
 Inpt:DATA Stats
 σ1:.49
 σ2:.33
 List1:L₁
 List2:L₂
 Freq1:0
↓Freq2:1
```

Select $>\mu_2$ as the alternative hypothesis and press **ENTER**. Scroll down to the next line and select **Calculate** or **Draw** and press **ENTER**.

The output for **Calculate** is:

```
2-SampZTest
 μ1>μ2
 z=4.975681069
 P=3.2559526ᴇ-7
 x̄1=2.13
 x̄2=1.593333333
↓Sx1=.490003519
```

The output for **Draw** is:

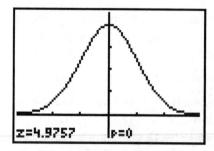

Since the P-value is smaller than α, the correct decision is to **Reject** H_o.

▶ Exercise 23 (pg. 377) Confidence Intervals for $\mu_1 - \mu_2$

Construct a confidence interval for $\mu_1 - \mu_2$, the difference between two population means. Designate the group using the herbal supplement as population 1 and the group using the placebo as population 2. The sample statistics for group 1 are: $\bar{x}_1 = 3.2$, $s_1 = 3.3$ and $n_1 = 42$. The sample statistics for group 2 are: $\bar{x}_2 = 4.1$, $s_2 = 3.9$ and $n_2 = 42$.

Press **STAT**, highlight **TESTS** and select **9:2-SampZInt** and press **ENTER**. Since you are using the sample statistics for your analysis, select **Stats** for **Inpt** and press **ENTER**. Enter the sample standard deviations as approximations for σ_1 and σ_2. Enter the sample means and sample sizes for each group.

```
2-SampZInt
 Inpt:Data Stats
 σ1:3.3
 σ2:3.9
 x1:3.2
 n1:42
 x2:4.1
↓n2:42
```

Scroll down to the next line and type in the confidence level of .95. Scroll down to the next line and press **ENTER**.

```
2-SampZInt
 (-2.445,.64505)
 x1=3.2
 x2=4.1
 n1=42
 n2=42
```

A 95 % confidence interval for the difference in the population means is (-2.445, .64505). Since this interval contains 0, the correct conclusion is that there is no difference in the means of the two groups. The herbal supplement does *not* appear to help in weight loss.

Section 8.2

▶ Example 1 (pg. 382) Two Sample t-Test

To test whether the mean stopping distances are different, use a two-tailed test: $H_o: \mu_1 = \mu_2$ vs $H_a: \mu_1 \neq \mu_2$. The sample statistics are found in the table at the top of pg. 382 in your textbook.

Press **STAT**, highlight **TESTS**, and select **4:2-SampTTest** and press **ENTER**. Since you are inputting the sample statistics, select **Stats** and press **ENTER**. Enter the sample information from the two samples. Select $\neq \mu_2$ as the alternative hypothesis and press **ENTER**. Scroll down to the next line. On this line, select **NO** because the variances are NOT assumed to be equal and therefore, you do not want a pooled estimate of the standard deviation. Press **ENTER**. Scroll down to the next line, highlight **Calculate** and press **ENTER**.

```
2-SampTTest
 μ1≠μ2
 t=-1.495915184
 P=.162535817
 df=11.10815972
 x̄1=51
↓x̄2=55
```

The output for **Calculate** displays the alternative hypothesis, the test statistic, the P-value, the degrees of freedom and the sample statistics. Notice the degrees of freedom = 11.108. In cases, such as this one, in which the population variances are not assumed to be equal, the calculator calculates an adjusted degrees of freedom, rather than using the smaller of (n_1-1) and (n_2-1).

If you choose **Draw**, the output includes a graph with the area associated with the P-value shaded.

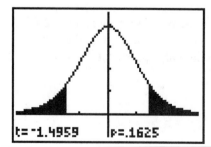

```
t=-1.4959    P=.1625
```

▶ Example 2 (pg. 383) Two Sample t-Test

To test the claim that the mean range of the manufacturer's cordless phone is greater than the mean range of the competitor's phone, use a right-tailed test with H_o: $\mu_1 \leq \mu_2$ vs H_a: $\mu_1 > \mu_2$. Designate the manufacturer's data as population 1 and the competitor's data as population 2. Use the Two-Sample T-Test for the analysis, and use the pooled variance option because the population variances are assumed to be equal in this example. The sample statistics are found in the table at the top of pg. 383 in your textbook.

Press **STAT**, highlight **TESTS**, and select **4:2-SampTTest** and press **ENTER**. Fill in the input screen with the sample statistics. Select $> \mu_2$ for the alternative hypothesis and press **ENTER**. Select **YES** for **Pooled** and press **ENTER**. To do the analysis, select **Calculate** and press **ENTER**.

```
2-SampTTest
 μ1>μ2
 t=1.811358919
 p=.0404131295
 df=28
 x̄1=1275
↓x̄2=1250
■
```

```
2-SampTTest
 μ1>μ2
↑Sx1=45
 Sx2=30
 SxP=37.7136769
 n1=14
 n2=16
■
```

The output display includes the alternative hypothesis, the test statistic, the P-value, degrees of freedom, the sample statistics and the pooled standard deviation (Sxp = 37.7)

If you select **Draw**, the display includes a graph with the shaded area associated with the test statistic, the P-value and the test statistic.

Since the P-value is less than α, the correct decision is to **Reject** H_o

◀

▶ Exercise 11 (pg. 385) Testing the Difference between 2 Means

Designate the small cars as population 1 and the midsize cars as population 2. Test the hypothesis: $H_o: \mu_1 = \mu_2$ vs. $H_a: \mu_1 \neq \mu_2$. The sample statistics are: $\bar{x}_1 = 23.1$, $s_1 = 8.69$, $n_1 = 14$, $\bar{x}_2 = 25.3$, $s_2 = 7.21$ and $n_2 = 23$. Use the Two-Sample T-test with the pooled variance option. Press **STAT**, highlight **TESTS** and select **4:2-SampTTest**. Choose **Stats** for **Inpt** and press **ENTER**. Enter the sample statistics. Select $\neq \mu_2$ as the alternative hypothesis and press **ENTER**. Select **YES** for **Pooled** and press **ENTER**. Highlight **Calculate** and press **ENTER**.

```
2-SampTTest
 µ1≠µ2
 t=-.8328515082
 p=.4105758669
 df=35
 x̄1=23.1
↓x̄2=25.3
■
```

Notice that the test statistic = -0.833, and the P-value is .411. Since the P-value is greater than α, the correct decision is to **Fail to Reject** H_o.

◀

▶ Exercise 19 (pg. 387) Testing the difference between 2 Means

Designate the "Old curriculum" as population 1 and the "New curriculum" as
population 2 and test the hypothesis: $H_o : \mu_1 \geq \mu_2$ vs. $H_a : \mu_1 < \mu_2$. Press
STAT, select 1:Edit and enter the data sets into L1 (Old curriculum) and L2
(New curriculum). Press STAT, highlight TESTS and select 4:2-SampTTest.
For this analysis, you are using the actual data so select Data for Inpt and press
ENTER. Fill in the input screen with the appropriate information. Choose $< \mu_2$
as the alternative hypothesis and press ENTER. Choose YES for Pooled and
press ENTER. Highlight Calculate and press ENTER.

```
2-SampTTest
 µ1<µ2
 t=-4.29519297
 P=5.0503069E-5
 df=42
 x̄1=56.68421053
↓x̄2=67.4
```

Since the P-value (.00005) is less than α, the correct decision is to Reject H_o.
This means that the data supports H_a, indicating that the new method of
teaching reading produces higher reading test scores than the old method. The
recommendation is to change to the new method.

◀

▶ Exercise 21 (pg. 388) Confidence Intervals

Compare the mean calorie content of grilled chicken sandwiches from Arby's restaurants to similar chicken sandwiches from McDonald's using a confidence interval. For this exercise, assume that the populations are normal and the variances are equal. Also, notice that the sample sizes are both less than 30. The appropriate confidence interval technique is the Two-Sample t-interval.

Press **STAT**, highlight **TESTS** and select **0:2-SampTInt**. For **Inpt**, select **Stats** and press **ENTER**. Enter the sample statistics from the two sets of data. Designate Arby's as population 1 and McDonald's as population 2. Choose .95 for **C-level** and choose **YES** for **Pooled** and press **ENTER**. Scroll down to the next line and press **ENTER** to **Calculate** the confidence interval.

```
2-SampTInt
 (-15.66,-4.337)
 df=25
 x̄1=230
 x̄2=240
 Sx1=6.2
↓Sx2=8.1
```

The output displays the 95 % confidence interval for (μ_1 - μ_2) and the sample statistics. One way to interpret the confidence interval is to state that "the difference in mean calorie content of grilled chicken sandwiches at the two restaurants is between -15.66 and -4.337." This means that the mean calorie content of the Arby's sandwiches is anywhere from 4.337 to 15.66 less than that of McDonald's.

▶ Exercise 23 (pg. 388) Confidence Intervals

Compare the mean cholesterol content of grilled sandwiches from Arby's to similar sandwiches from McDonald's using a confidence interval. For this exercise, assume the populations are normal but do NOT assume equal variances. Use the sample sizes from Exercise 21.

Press **STAT**, highlight **TESTS** and select **0:2-SampTInt**. For **Inpt**, select **Stats** and press **ENTER**. Enter the sample statistics from the table in Exercise 23. Designate Arby's as population 1 and McDonald's as population 2. Choose **.90** for **C-level** and choose **NO** for **Pooled** and press **ENTER**. Scroll down to the next line and press **ENTER** to **Calculate** the confidence interval.

```
2-SampTInt
 (-.9816,2.9816)
 df=24.37225626
 x̄1=61
 x̄2=60
 Sx1=3.59
↓Sx2=2.41
```

The output displays the 90 % confidence interval for ($\mu_1 - \mu_2$), the adjusted degrees of freedom and the sample statistics. The confidence interval states that the mean cholesterol content for Arby's sandwiches is anywhere from 0.98 less than to 2.98 more than that of McDonald's. Since this interval contains 0, you cannot conclude that there is any difference in mean cholesterol content.

◀

Section 8.3

▶ Example 2 (pg. 392) The Paired t-Test

In this example, the data is paired data, with two scores for each of the 8 golfers. Enter the scores that each golfer gave as his or her most recent score into L1. Enter the scores achieved after using the newly designated clubs into L2. Next, you must create a set of differences, d = (old score) - (new score). To create this set, move the cursor to highlight the label **L3** and press **ENTER**. Notice that the cursor is flashing on the bottom line of the display. Press 2nd **[L1]** - 2nd **[L2]**

L1	L2	▨	3
89	83	------	
84	83		
96	92		
82	84		
74	76		
92	91		
85	80		

L3 =L₁−L₂

and press **ENTER**.

L1	L2	L3	3
89	83	6	
84	83	1	
96	92	4	
82	84	-2	
74	76	-2	
92	91	1	
85	80	5	

L3(1)=6

Each value in **L3** is the difference **L1 - L2**.

To test the claim that golfers can lower their scores using the manufacturer's newly designed clubs, the hypothesis test is: $H_o: \mu_d \leq 0$ vs. $H_a: \mu_d > 0$. Press **STAT**, highlight **TESTS** and select **2:T-Test**. In this example, you are using the actual data to do the analysis, so select **Data** for **Inpt** and press **ENTER**. The value for μ_o is **0**, the value in the null hypothesis. The set of differences is found in **L3**, so set **List** to **L3**. Set **Freq** equal to **1**. Choose $> \mu_o$ as the alternative hypothesis and highlight **Calculate** and press **ENTER**.

```
T-Test
 μ>0
 t=1.498259585
 P=.0888692418
 x̄=1.625
 Sx=3.067688753
 n=8
```

You can also highlight **DRAW** and press **ENTER**.

Since the P-value is less than α, the correct decision is to **Reject** H_o.

◀

▶ Exercise 15 (pg. 396) Paired Difference Test

Press **STAT** and select **1:Edit.** Enter the students' scores on the first SAT into **L1** and their scores on the second Sat into **L2**. Since this is paired data, create a column of differences, d = (first SAT) - (second SAT). To create the differences, move the cursor to highlight the label **L3** and press **ENTER**. Notice that the cursor is flashing on the bottom line of the display. Press **2ⁿᵈ** **[L1]** - **2ⁿᵈ** **[L2]** and press **ENTER**.

```
L1        L2        L3        3
 445       446       -1
 510       571       -61
 429       517       -88
 452       478       -26
 629       610       19
 433       453       -20
 551       516       35
L3(1)= -1
```

Each value in **L3** is the difference **L1 - L2**.

To test the claim that students' verbal SAT scores improved the second time they took the verbal SAT, the hypotheses are:: $H_o: \mu_d \geq 0$ vs. $H_a: \mu_d < 0$. Press **STAT**, highlight **TESTS** and select **2:T-Test**. In this example, you are using the actual data to do the analysis, so select **Data** for **Inpt** and press **ENTER**. The value for μ_o is **0**, the value in the null hypothesis. The set of differences is found in **L3**, so set **List** to **L3**. Set **Freq** equal to **1**. Choose < μ_o as the alternative hypothesis and highlight **Calculate** and press **ENTER**.

```
T-Test
 μ<0
 t=-3.001096523
 p=.0051086649
 x̄=-33.71428571
 Sx=42.03373841
 n=14
```

Or, highlight **DRAW** and press **ENTER**.

Since the P-value is less than α, the correct decision is to **Reject** H_o.

◀

> ▸ Exercise 23 (pg. 399)   Confidence Interval for μ_d

Press **STAT** and select **1:Edit**. Enter the data for "hours of sleep without the drug" into **L1** and the data for "hours of sleep using the new drug" into **L2**. Since this is paired data, create a column of differences, **d = L1 - L2**. To create the differences, move the cursor to highlight the label **L3** and press **ENTER**. Notice that the cursor is flashing on the bottom line of the display. Press **2ⁿᵈ [L1]** - **2ⁿᵈ [L2]** and press **ENTER**. Each value in **L3** is the difference **L1 - L2**.

To construct a confidence interval for μ_d, press **STAT**, highlight **TESTS** and select **8:Tinterval**. For **Inpt**, select **Data** and press **ENTER**. Fill in the appropriate information and highlight **Calculate**.

```
TInterval
 (-1.762,-1.171)
 x̄=-1.466666667
 Sx=.569422887
 n=12
```

The confidence interval for μ_d is -1.762 to -1.171. This means that the average difference in hours of sleep is 1.171 hours to 1.762 hours more for patients using the new drug.

Section 8.4

▶ Example 1 (pg. 402) Testing the Difference Between p_1 and p_2

To test the claim that there is a difference in the proportion of female Internet users who plan to shop On-line and the proportion of male Internet users who plan to shop On-line, the correct hypothesis test is: $H_o: p_1 = p_2$ vs $H_a: p_1 \neq p_2$. Designate the females as population 1 and the males as population 2. The sample statistics are $n_1 = 200$, $p_1 = .30$, $n_2 = 250$, and $p_2 = .38$. To calculate x_1, the number of females in the sample who plan to shop On-line, multiply n_1 times p_1. To calculate x_2, multiply n_2 times p_2.

Press **STAT**, highlight **TESTS** and select **6:2-PropZTest** and fill in the appropriate information. Highlight **Calculate** and press **ENTER**.

```
2-PropZTest
 P1≠P2
 z=-1.774615984
 P=.0759612188
 P1=.3
 P2=.38
↓P=.344444444
```

The output displays the alternative hypothesis, the test statistic, the P-value, the sample statistics and the weighted estimate of the population proportion, $\hat{p}$. Since the P-value is less than α, the correct decision is to **Reject** H_o.

◀

▶ Exercise 7 (pg. 404) The Difference Between Two Proportions

Designate the 1990 data as population 1 and the more recent data as population 2 and test the hypotheses: $H_o: p_1 = p_2$ vs $H_a: p_1 \neq p_2$. The sample statistics are $n_1 = 1539$, $x_1 = 520$, $n_2 = 2055$, and $x_2 = 865$.

Press **STAT**, highlight **TESTS** and select **6:2-PropZTest** and fill in the appropriate information. Highlight **Calculate** and press **ENTER**.

```
2-PropZTest
 P1≠P2
 z=-5.06166817
 P=4.1630154E-7
 p1=.3378817414
 p2=.4209245742
↓p=.3853644964
■
```

Since the P-value (.0000004163) is less than α, the correct decision is to **Reject** H_o. This indicates that the proportion of adults using alternative medicines has changed since 1990.

◀

▶ Exercise 17 (pg. 406)   Confidence Interval for $p_1 - p_2$.

Construct a confidence interval to compare the proportion of students who had planned to study engineering several years ago to the proportion currently planning on studying engineering. Designate the earlier survey results as population 1 and the recent survey results as population 2. The sample statistics are n_1 = 977000, p_1 = 11.7 %, n_2 = 1085000, and p_2 =8.8 %. To calculate x_1 multiply n_1 times p_1. (Use .117 for p_1 and .088 for p_2.) To calculate x_2, multiply n_2 times p_2.

Press **STAT**, highlight **TESTS** and select **B:2-PropZInt** and fill in the appropriate information. Highlight **Calculate** and press **ENTER**.

```
2-PropZInt
 (.02817,.02983)
 p̂1=.117
 p̂2=.088
 n1=977000
 n2=1085000
```

The confidence interval (.02817, .02983) indicates that the proportion of students having chosen engineering in the past is between 2.817 % and 2.983 % higher than the proportion of students currently choosing engineering. Notice how narrow the confidence interval is. This is due to the very large sample sizes.

◀

▸ Technology Lab (pg. 407) Tails Over Heads

Exercise 1 - 2: Test the hypotheses: H_o : P(Heads) = .5 vs. H_a : P(Heads) ≠ .5 using the one sample test of a proportion. Press **STAT**, highlight **TESTS** and select **5:1-PropZTest.** For this example, p_o = .5. Using Casey's data, X = 5772 and n = 11902. The alternative hypothesis is ≠ . Highlight **Calculate** and press **ENTER**.

```
1-PropZTest
 Prop≠.5
 z=-3.281504874
 P=.0010326665
 p̂=.4849605108
 n=11902
■
```

Since the P-value is less than α , the correct decision is to **Reject** H_o .

Exercise 3: The histogram at the top of the page is a graph of 500 simulations of Casey's experiment. Each simulation represents 11902 flips of a fair coin. The bars of the histogram represent frequencies. Use the histogram to determine how often 5772 or fewer heads occurred.

To simulate this experiment, you must use an alternative technology. The TI-83 does not have the memory capacity to do this experiment.

Exercise 4: To compare the mint dates of the coins, run the hypothesis test: H_o : μ_1 = μ_2 vs. H_a : μ_1 ≠ μ_2 . Designate the Philadelphia data as population 1 and the Denver data as population 2.

Press **STAT**, highlight **TESTS** and select **3:2-SampZTest**. Choose **Stats** for **Inpt** and press **ENTER**. Enter the sample statistics. Use the sample standard deviations as approximations to the population standard deviations. Select ≠ μ_2 as the alternative hypothesis and press **ENTER**. Highlight **Calculate** and press **ENTER**.

```
2-SampZTest
 µ1≠µ2
 z=8.801919011
 P=1.363493E-18
 x̄1=1984.8
 x̄2=1983.4
↓n1=7133
```

Since the P-value is extremely small, the correct decision is to **Reject** H_o.

Exercise 5: To compare the average mint value of coins minted in Philadelphia to those minted in Denver, run the hypothesis test: $H_o : \mu_1 = \mu_2$ vs. $H_a : \mu_1 \neq \mu_2$. Designate the Philadelphia data as population 1 and the Denver data as population 2.

Press STAT, highlight **TESTS** and select **3:2-SampZTest**. Choose **Stats** for **Inpt** and press ENTER. Enter the sample statistics. Use the sample standard deviations as approximations to the population standard deviations. Select $\neq \mu_2$ as the alternative hypothesis and press ENTER. Highlight **Calculate** and press ENTER. Since the P-value is extremely small, the correct decision is to **Reject** H_o.

Correlation and Regression

CHAPTER

9

Section 9.1

▶ Example 3 (pg. 420) Constructing a Scatter plot

Press **STAT**, highlight **1:Edit** and clear **L1** and **L2**. Enter the X-values into **L1** and the Y-values into **L2**. Press **2ⁿᵈ** **STAT PLOT]** , select **1:Plot1**, turn **ON** Plot 1 and press **ENTER**. For **Type** of graph, select the **scatter plot** which is the first selection. Press **ENTER**. Enter **L1** for **Xlist** and **L2** for **Ylist**. Highlight the first selection, the small square, for the type of **Mark**. Press **ENTER**. Press **ZOOM** and **9** to select **ZoomStat**.

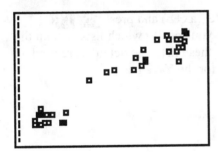

This graph shows a positive linear correlation. (Note: if you would like the X and Y axes to start at 0, press **WINDOW** and set **Xmin = 0** and **Ymin = 0**. The picture will then look like the one in your textbook.)

▶ Example 5 (pg. 423)   Finding a Correlation Coefficient

For this example, use the data from Example 3 on pg. 420. Enter the X-values into **L1** and the Y-values into **L2**. In order to calculate r, the correlation coefficient, you must turn **On** the **Diagnostic** command. Press **2ⁿᵈ [CATALOG]** (Note: CATALOG is found above the ⓪ key). The CATALOG of functions will appear on the screen. Use the down arrow to scroll to the **DiagnosticOn** command.

Press **ENTER ENTER**.

Press **STAT**, highlight **CALC**, select **4:LinReg(ax+b)** and press **ENTER ENTER**. (Note: This command requires that you specify which lists contain the X-values and Y-values. If you do not specify these lists, the defaults are used. The defaults are: **L1** for the X-values and **L2** for the Y-values.)

```
LinReg
 y=ax+b
 a=11.8244078
 b=35.30117105
 r²=.9404868083
 r=.9697869912
```

The correlation coefficient is r = .9697869912. This suggests a strong positive linear correlation between X and Y.

◀

▶ Exercise 13 (pg. 428) Constructing a Scatter plot and
                        Determining r

Enter the X-values into **L1** and the Y-values into **L2**. Press **2ⁿᵈ** **[STAT PLOT]** ,
select **1:Plot1**, turn **ON** Plot 1 and press **ENTER**. For **Type** of graph, select the
scatter plot. Enter **L1** for **Xlist** and **L2** for **Ylist**. Press **ZOOM** and **9** to select
ZoomStat.

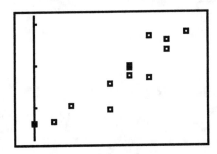

To calculate r, press **STAT**, highlight **CALC** and select **4:LinReg(ax+b)** and
press **ENTER** **ENTER**.

```
LinReg
 y=ax+b
 a=7.349665924
 b=34.6169265
 r²=.8512827753
 r=.922649866
```

The scatter plot shows a strong positive linear correlation. This is confirmed by
the r-value of 0.923.

◀

> ▶ Example 27 (pg. 430)   Testing Claims

To test the significance of the population correlation coefficient, ρ, the appropriate hypothesis test is: $\rho = 0$ vs. $\rho \neq 0$. To run the test, enter the X-values into **L1** and the Y-values into **L2**. Press **STAT**, highlight **TESTS** and select **E:LinRegTTest**. Enter **L1** for **Xlist**, **L2** for **Ylist**, and **1** for **Freq**. On the next line, β and ρ, select $\neq 0$ and press **ENTER**. Leave the next line, RegEQ, blank. Highlight **Calculate.**

```
LinRegTTest
 Xlist:L₁
 Ylist:L₂
 Freq:1
 β & ρ:≠0 <0 >0
 RegEQ:
 Calculate
```

Press **ENTER**.

```
LinRegTTest
 y=a+bx
 β≠0 and ρ≠0
 t=7.935104115
 P=7.0576131E-6
 df=11
↓a=34.6169265
```

The output displays several pieces of information describing the relationship between X and Y. What you are interested in for this example are the following: the test statistic (t = 7.935), the P-value (p = 7.0576131E-6) and the r value (r = .922649886). Since the P-value is less than α, the correct decision is to **Reject** the null hypothesis. This indicates that there is a linear relationship between X and Y.

Section 9.2

▶ Example 2 (pg. 434)   Finding a Regression Line

Enter the X-values into **L1** and the Y-values into **L2**. Press **STAT**, highlight **CALC** and select **4:LinReg(ax+b)**. This command has several options. One option allows you to store the regression equation into one of the Y-variables. To use this option, with the cursor flashing on the line **LinReg(ax+b)**, press **VARS**.

Highlight **Y-VARS.**

Select **1:Function** and press **ENTER**

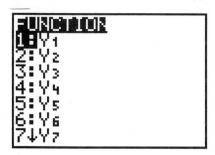

Notice that **1:Y1** is highlighted. Press **ENTER**.

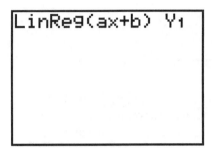

```
LinReg(ax+b) Y₁
```

Press **ENTER**.

```
LinReg
 y=ax+b
 a=11.8244078
 b=35.30117105
 r²=.9404868083
 r=.9697869912
```

The output displays the general form of the regression equation: $y = ax+b$
followed by values for a and b. Next, r^2, the coefficient of determination, and r,
the correlation coefficient , are displayed. If you put the values of a and b into
the general equation, you obtain the specific linear equation for this data:
$y = 11.82 + 35.30\ x$. Press **Y=** and see that this specific equation has been pasted
to **Y1**.

```
Plot1  Plot2  Plot3
\Y₁◻11.824407800
243X+35.30117104
6452
\Y₂=
\Y₃=
\Y₄=
\Y₅=
```

Press **2ⁿᵈ STAT PLOT]** , select **1:Plot1**, turn **ON** Plot1, select **scatter plot**, set
Xlist to **L1** and **Ylist** to **L2**. Press **ZOOM** and **9**.

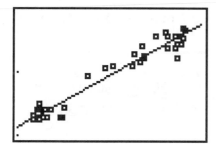

This picture displays a scatter plot of the data and the regression line. The picture indicates a strong positive linear correlation between X and Y, which is confirmed by the r-value of .970.

You can use the regression equation stored in **Y1** to predict Y-values for specific X-values. For example, suppose the duration of an eruption was equal to 1.95 minutes. Predict the time (in minutes) until the next eruption. In other words, for X = 1.95, what does the regression equation predict for Y? To find this value for Y, press **VARS**, highlight **Y-VARS**, select **1:Function**, press **ENTER**, select **1:Y1** and press **ENTER**. Press **(** 1.95 **)** and press **ENTER** .

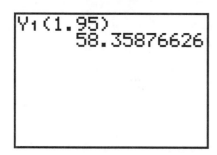

The output is a display of the predicted Y-value for X = 1.95.

▶ Exercise 11 (pg. 437)   Finding the Equation of the Regression
Line

Enter the X-values into **L1** and the Y-values into **L2**. Press **STAT**. Highlight
CALC, select **4:LinReg(ax+b)**, press **ENTER**. Press **VARS**, highlight **Y-
VARS**, select **1:Function**, press **ENTER** and select **1:Y1** and press **ENTER**.

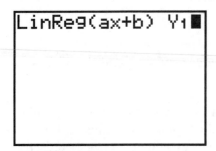

Press **ENTER**.

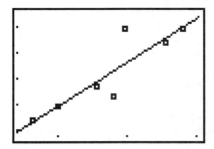

Using **a** and **b** from the output display, the resulting regression equation is y =
1.724 x + 79.73. Press **Y=** to confirm that the regression equation has been
stored in **Y1**. Press **ZOOM** and **9** for **ZoomStat** and a graph of the scatter plot
with the regression line will be displayed.

Next, you can use the regression equation to predict systolic blood pressure, Y,
for various ages, X. First, check the X-values that you will be using to confirm

that they are within (or close to) the range of the X-values in your data. The four X-values (18, 71, 29 and 55) meet this criteria.

Press **VARS**, highlight **Y-VARS**, select **1:Function** and press **ENTER**. Select **1:Y1** and press **ENTER**. Press **(** 18 **)** and **ENTER**.

```
Y₁(18)
       110.7580483



```

The predicted systolic blood pressure for the average 18-year-old male is 110.8.

Press **2ⁿᵈ** [ENTRY], (found above the ENTER key). Move the cursor so that it is flashing on **1** and type in 71. Press **ENTER**.

```
Y₁(18)
       110.7580483
Y₁(71)
       202.1084004
■

```

The predicted systolic blood pressure for the average 71-year-old male is 202.1.

Press **2ⁿᵈ** [ENTRY] . Move the cursor so that it is flashing on **7** and type in 29. Press **ENTER**.

The predicted systolic blood pressure for the average 29-year-old male is 129.7.

Press **2ⁿᵈ** [ENTRY] . Move the cursor so that it is flashing on **2** and type in 55. Press **ENTER**. The predicted systolic blood pressure for the average 55-year-old male is 174.55.

◀

Section 9.3

▶ Example 2 (pg. 445)   The Standard Error of the Estimate

Enter the data into **L1** and **L2**. Press **STAT**, highlight **CALC**, select
4:LinReg(ax+b), and press **ENTER**. Press **VARS**, highlight **Y-VARS**, select
1:Function, press **ENTER**, select **1:Y1** and press **ENTER** **ENTER**.

The formula for s_e, the standard error of the estimate is $\sqrt{\dfrac{\sum (y_i - \hat{y}_i)^2}{n-2}}$. The
values for $(y_i - \hat{y}_i)$, called Residuals, are automatically stored to a list called
RESID. Press **2ⁿᵈ** [LIST] , select **RESID** and press **ENTER**. Press **STO**, **2ⁿᵈ**
[L3] and **ENTER**. This stores the residuals to **L3**.

```
.RESID→L₃
{-.8097165992  -...
```

In the formula for s_e, the residuals, $(y_i - \hat{y}_i)$, are squared. To square these
values and store them in **L4**, press **STAT**, select **1:Edit** and move the cursor to
highlight the Listname **L4**. Press **ENTER**. Press **2ⁿᵈ** [L3] and the x^2 key.

```
L2      L3      L4       4
225     -.8097  ------
184     -1.227
220     14.482
240     4.0445
180     4.919
184     -1.227
186     -19.52
L4 =L₃²
```

Press **ENTER**.

```
L2      L3       L4        4
225     -.8097   .65564
184     -1.227   1.5048
220     14.482   209.72
240     4.0445   16.358
180     4.919    24.197
184     -1.227   1.5048
186     -19.52   380.96
L4(1)=.6556409710...
```

Press **2**nd **[QUIT]** . Next, you need to find the sum of **L4**, $\sum (y_i - \hat{y}_i)^2$. Press **2**nd **[LIST]** . Highlight **MATH**, select **5:sum(** and press **2**nd **[L4]** . Close the parentheses and press **ENTER**.

```
sum(L4)
        635.3441296
```

Divide this sum by (number of observations – 2). In this example, (n - 2) = 6. Simply press ÷ 6 and press **ENTER**.

```
sum(L4)
        635.3441296
Ans/6
        105.8906883
```

Lastly, take the square root of this answer by pressing **2**nd **[√]** **2**nd **[ANS]** . Close the parentheses and press **ENTER**.

```
sum(L4)
        635.3441296
Ans/6
        105.8906883
√(Ans)
        10.29032012
█
```

The standard error of the estimate, s_e, is 10.29.

▶ Example 3 (pg. 447)   Constructing a Prediction Interval

This example is a continuation of Example 2 on pg. 445. To construct a prediction interval for a specific X-value, x_o, you must calculate the maximum

error, E. The formula for E is: $E = t_c s_e \sqrt{(1 + \dfrac{1}{n} + \dfrac{n(x_o - \bar{x})^2}{n(\sum x^2) - (\sum x)^2})}$. The

critical value for t is found in the t-table. For this example, $t_c = 2.447$. The standard error, s_e, is 10.290.

To calculate $\bar{x}$, $\sum x^2$, and $(\sum x)^2$, press **VARS**, select **5:Statistics** and press **ENTER**. Highlight **2:$\bar{x}$** and press **ENTER ENTER**. Notice that $\bar{x} = 1.975$. Press **VARS** again, select **5:Statistics**, highlight $\sum$, select **1:$\sum x$** and press **ENTER ENTER**. So, $\sum x = 15.8$. Press **VARS** again, select **5:Statistics**, highlight $\sum$, select **2: $\sum x^2$** , and press **ENTER ENTER**. Notice that $\sum x^2 = 32.44$.

Next, calculate the maximum error, E, when $x_o = 2.1$ using the formula for E.

```
2.447*10.29*√(1+
1/8+8(2.1-1.975)
²/(8*32.44-15.8²
)
        26.85678531
```

The prediction interval is $\hat{y} \pm 26.857$. To find $\hat{y}$, the predicted value for y when x = 2.1, press **VARS**, highlight **Y-VARS**, select **1:Function** and press **ENTER**. Next select **1:Y1**, press **ENTER** and press **(** 2.1 **)**. Finally, press **ENTER**.

```
2.447*10.29*√(1+
1/8+8(2.1-1.975)
²/(8*32.44-15.8²
)
        26.85678531
Y₁(2.1)
        210.5910931
■
```

The prediction interval is 210.59 ± 26.857.

▶ Exercise 11 (pg. 449)   Coefficient of Determination and
Standard Error of the Estimate

Enter the data into **L1** and **L2**. Press **STAT**, highlight **CALC**, select
4:LinReg(ax+b) and press **ENTER**. Next, press **VARS**, select **Y-VARS**, select
1:Function and press **ENTER**. Lastly, select **1:Y1** and press **ENTER** **ENTER**.

```
LinReg
 y=ax+b
 a=230.8314924
 b=-289.8094201
 r²=.9853951741
 r=.9926707279
█
```

The coefficient of determination, r^2, is .9854. This means that 98.54 % of the
variation of the Y-values is explained by the X-values.

To calculate s_e, press 2nd [LIST] , select **1:RESID**, press **ENTER**, **STO** , 2nd
[L3] **ENTER**. This stores the residuals to **L3**. In the formula for s_e, the
residuals, $(y_i - \hat{y}_i)$, are squared. To square these values and store them in **L4**,
press **STAT**, select **1:Edit** and move the cursor to highlight the Listname **L4**.
Press **ENTER**. Press 2nd [L3] and the x^2 key. Press **ENTER**.

```
L2       L3       L4       4
123.2    43.679   1907.9
211.5    -29.6    876.34
385.5    -17.19   295.33
475.1    -19.92   396.71
641.1    30.667   940.44
716.9    -55.12   3037.7
768.2    -26.9    723.53
L4(1)=1907.857859...
```

Press 2nd [QUIT] . Next, you need to find the sum of **L4**, $\sum (y_i - \hat{y}_i)^2$. Press
2nd [LIST] . Highlight **MATH**, select **5:sum(** and press 2nd [L4] . Close the
parentheses and press **ENTER**.

```
 r²=.9853951741
 r=.9926707279

∟RESID→L₃
{43.67903226  -2...
sum(L₄)
       11439.53051
█
```

Divide this sum by (number of observations − 2). In this example, (n - 2) = 9.
Simply press ÷ 9 and press ENTER.

```
∟RESID→L₃
{43.67903226  -2...
sum(L₄)
       11439.53051
Ans/9
       1271.058945
█
```

Lastly, take the square root of this answer by pressing 2ⁿᵈ [√] 2ⁿᵈ [ANS].
Close the parentheses and press ENTER.

```
{43.67903226  -2...
sum(L₄)
       11439.53051
Ans/9
       1271.058945
√(Ans)
       35.65191363
█
```

▶ Technology Lab (pg. 457) Tar, Nicotine and Carbon Monoxide

Exercises 1-2: Enter the data into **L1**, **L2**, **L3** and **L4**. To construct the scatter plots, press **2ⁿᵈ** [STAT PLOT], select **1:Plot1** and press ENTER. Turn **ON** Plot1. Select **scatter plot** for **Type**. Enter the appropriate labels for **Xlist** and **Ylist** to construct each of the scatter plots.

Exercises 3: To find the correlation coefficients, press STAT, highlight **CALC** and select **4:LinReg(ax+b)** and enter the labels of the columns you are using for the correlation. For example, to find the correlation coefficient for "weight" and "carbon monoxide", press **2ⁿᵈ** [L3] , **2ⁿᵈ** [L4].

```
LinReg(ax+b) L₃,
L₄
```

Press ENTER and notice that the correlation coefficient for weight and carbon monoxide is .3616.

Exercise 4: To find the regression equations, press STAT, highlight **CALC** and select **4:LinReg(ax+b)** and enter the labels of the columns you are using for the regression. For example, to find the regression equation for "tar" and "carbon monoxide", press **2ⁿᵈ** [L1] , **2ⁿᵈ** [L4] .

Exercise 5: Use the regression equations found in Exercise 4 to do the predictions.

◀

Chi-Square Tests and the F-Distribution

CHAPTER

10

Section 10.2

▶ Example 3 (pg. 481)   Chi-Square Independence Test

Test the hypotheses: H_o :The number of days spent exercising per week is *independent* of gender vs. H_a : The number of days spent exercising per week *depends* on gender. Enter the data in the table into **Matrix A**. Press MATRX, highlight **EDIT** and press ENTER.

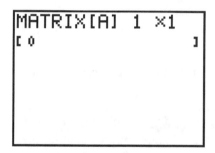

On the top row of the display, enter the size of the matrix. The matrix has 2 rows and 4 columns, so press 2 , press the right arrow key, and press 4. Press ENTER. Enter the first value, 40, and press ENTER. Enter the second value, 53, and press ENTER. Continue this process and fill the matrix.

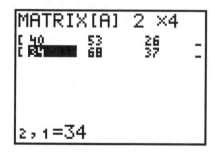

Press 2nd [Quit] . To perform the test of independence, press STAT, highlight **TESTS**, and select **C:** χ^2**-Test** and press ENTER.

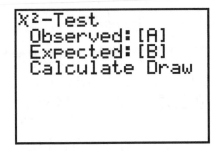

For **Observed**, [A] should be selected. If [A] is not already selected, press
MATRX, highlight **NAMES**, select **1:[A]** and press ENTER. For, **Expected**, [B]
should be selected. Move the cursor to the next line and select **Calculate** and
press ENTER.

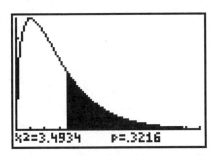

The output displays the test statistic and the P-value. Since p is greater than α,
the correct decision is to **Fail to Reject** the null hypothesis. This means that the
number of days per week spent exercising is *independent* of gender.

Or, you could highlight **Draw** and press ENTER.

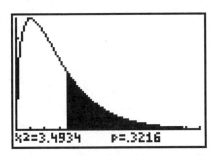

This output displays the χ^2 –**curve** with the area associated with the P-value
shaded in. The test statistic and the P-value are also displayed.

▶ Exercise 5 (pg. 483)   Chi-Square Test for Independence

Test the hypotheses: H_o : The Result (improvement or no change) is *independent* of Treatment (drug or placebo) vs. H_a : The Result *depends* on the Treatment. Enter the data in the table into **Matrix A**. Press MATRX, highlight **EDIT** and press ENTER.

On the top row of the display, enter the size of the matrix. The matrix has 2 rows and 2 columns, so press 2 , press the right arrow key, and press 2. Press ENTER. Enter the first value, 39, and press ENTER. Enter the second value, 25, and press ENTER. Continue this process and fill the matrix.

Press 2nd [Quit] . Press STAT, highlight **TESTS**, and select **C: χ^2-Test** and press ENTER. For **Observed**, **[A]** should be selected. If **[A]** is not already selected, press MATRX, highlight **NAMES**, select **1:[A]** and press ENTER. For, **Expected**, **[B]** should be selected. Move the cursor to the next line and select **Calculate** and press ENTER.

```
X²-Test
 X²=5.106317432
 P=.0238388683
 df=1
```

The output displays the test statistic and the P-value. Since p is less than α, the correct decision is to **Reject** the null hypothesis. The correct recommendation is to use the drug as part of the treatment.

Section 10.3

▶ Example 3 (pg. 491)   Performing a Two-Sample F-Test

Test the hypotheses: $H_o : \sigma_1^2 \leq \sigma_2^2$ vs. $H_a : \sigma_1^2 > \sigma_2^2$. The sample statistics are: $s_1^2 = 144$, $n_1 = 10$, $s_2^2 = 100$ and $n_2 = 21$. Press STAT, highlight **TESTS** and select **D:2-SampFTest**. For **Inpt**, select **Stats** and press ENTER. On the next line, enter s_1, the standard deviation under the old system. The standard deviation is the square root of the variance, so press 2nd [$\sqrt{\ }$] and enter 144. Enter n_1 on the next line. Next, enter the standard deviation under the new system by pressing 2nd [$\sqrt{\ }$] 100. Enter n_2 on the next line. Highlight $> \sigma_2$ for the alternative hypothesis and press ENTER. Select **Calculate** and press ENTER.

```
2-SampFTest
 σ1>σ2
 F=1.44
 P=.2369000822
 Sx1=12
 Sx2=10
↓n1=10
```

The output displays the alternative hypothesis, the test statistic, the P-value and the sample statistics. Since the P-value is greater than α, the correct decision is to **Fail to Reject** the null hypothesis. There is not enough evidence to conclude that the new system decreased the variance in waiting times.

You could also select **Draw** and press ENTER.

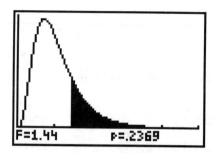

The output displays the F-distribution with the area associated with the P-value shaded in.

▶ Exercise 15 (pg. 494)   Comparing Two Variances

Test the hypotheses: $H_o : \sigma_1^2 \le \sigma_2^2$ vs. $H_a : \sigma_1^2 > \sigma_2^2$. The sample statistics are: $s_1 = 0.7$, $n_1 = 25$, $s_2 = 0.5$ and $n_2 = 21$. Press STAT, highlight **TESTS** and select **D:2-SampFTest**. For **Inpt**, select **Stats** and press ENTER. On the next line, enter s_1, the standard deviation under the old procedure. Enter n_1 on the next line. Next, enter the standard deviation and sample size for the new procedure and press ENTER. Highlight $> \sigma_2$ for the alternative hypothesis and press ENTER.   Select **Calculate** and press ENTER.

```
2-SampFTest
 σ1>σ2
 F=1.96
 P=.0653066513
 Sx1=.7
 Sx2=.5
↓n1=25
```

The output displays the alternative hypothesis, the test statistic, the P-value and the sample statistics. Since the P-value is less than α, the correct decision is to **Reject** the null hypothesis. The data supports the hospital's claim that the standard deviation of the waiting times has decreased.

◀

Section 10.4

▶ Example 2 (pg. 501)   Performing an ANOVA Test

Test the hypotheses: $H_o : \mu_1 = \mu_2 = \mu_3$ vs. H_a :at least one mean is different from the others. Enter the data into **L1, L2,** and **L3.** Press **STAT**, highlight **TESTS** and select **F:ANOVA(** and press **ENTER**, 2nd [L1] , 2nd [L2] , 2nd [L3].

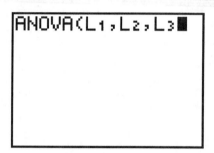

Press **ENTER** and the results will be displayed on the screen.

```
One-way ANOVA
 F=6.13023478
 p=.0063828069
 Factor
  df=2
  SS=1806.46667
↓ MS=903.233333
■
```

```
One-way ANOVA
↑ MS=903.233333
 Error
  df=27
  SS=3978.2
  MS=147.340741
 Sxp=12.1383994
■
```

The output displays the test statistic, F = 6.13, and the P-value, p = .00638. Since the P-value is less than α, the correct decision is to **Reject** the null hypothesis.

This indicates that there is a difference in the mean flight times of the three airlines.

The output also displays all the information that is needed to set up an ANOVA table like the table on pg. 498 of your textbook.

Variation	Sum of Squares	Degrees of Freedom	Mean Squares	F
Between	1806.5	2	903.2	903.2 / 147.3
Within	3978.2	27	147.3	

Note: The TI-83 labels variation "Between" samples as variation due to the "Factor". Also, variation "Within" samples is labeled as "Error".

The final item in the output is the pooled standard deviation, Sxp = 12.14.

▶ Exercise 1 (pg. 503)   Performing an ANOVA Test

Test the hypotheses: $H_o : \mu_1 = \mu_2 = \mu_3$ vs. H_a :at least one mean is different from the others. Enter the data into **L1, L2**, and **L3**. Press STAT, highlight **TESTS** and select **F:ANOVA(** and press ENTER, 2nd [L1] , 2nd [L2] , 2nd [L3]. Press ENTER and the results will be displayed on the screen.

```
One-way ANOVA
 F=1.259699793
 P=.2998881447
 Factor
  df=2
  SS=1.63002551
↓ MS=.815012756
```

```
One-way ANOVA
↑ MS=.815012756
 Error
  df=27
  SS=17.4687212
  MS=.646989672
 Sxp=.804356682
```

The output displays the test statistic, F = 1.26, and the P-value, p = 0.299888. Since the P-value is greater than α, the correct decision is to **Fail to Reject** the null hypothesis. The data does not support the claim that at least one mean cost per month is different.

◀

▶ Exercise 9 (pg. 506)   Performing an ANOVA Test

Test the hypotheses: $H_o : \mu_1 = \mu_2 = \mu_3 = \mu_4$ vs. H_a :at least one mean is different from the others. Enter the data into **L1, L2, L3** and **L4** . Press **STAT**, highlight **TESTS** and select **F:ANOVA(** and press **ENTER**, 2nd [L1] , 2nd [L2] , 2nd [L3] , 2nd [L4]. Press **ENTER** and the results will be displayed on the screen.

```
One-way ANOVA
 F=8.4596944
 p=2.1962975E-4
 Factor
  df=3
  SS=61130.6776
↓ MS=20376.8925
```

```
One-way ANOVA
↑ MS=20376.8925
 Error
  df=36
  SS=86713.3134
  MS=2408.70315
 Sxp=49.0785406
█
```

The output displays the test statistic, F = 8.46, and the P-value, p = .0002196. Since the P-value is less than α , the correct decision is to **Reject** the null hypothesis. The data supports the claim that the mean energy consumption for at least one region is different.

◀

▸ Technology Lab (pg. 508)   Crash Tests

Exercise 2: Test the hypotheses: $H_o \, \sigma_1^{\;2} = \sigma_2^{\;2}$ vs. $H_a : \sigma_1^{\;2} \neq \sigma_2^2$ for each pair of samples. This test requires values for s_1 and s_2. To obtain these values, enter the data into **L1, L2** and **L3**. To find the standard deviation for **L1**, press STAT, highlight **CALC** and press ENTER ENTER. To find the standard deviation for **L2**, press STAT, highlight **CALC**, press ENTER, 2nd [L2] ENTER. Repeat this process to get the standard deviation of **L3**.

For each comparison, press STAT, highlight **TESTS** and select **D:2-SampFTest**. For **Inpt**, select **Stats** and press ENTER. On the next line, enter s_1, the larger of the two standard deviations. Enter n_1 on the next line. Next, enter the smaller of the two standard deviations and then, the sample size. Highlight $\neq \sigma_2^{\;2}$ for the alternative hypothesis and press ENTER. Select **Calculate** and press ENTER.

Exercise 3: One method of testing to see if a set of data is approximately normal is to use a Normal Probability Plot. To test the data in **L1**, press 2nd [STAT PLOT] , select Plot 1 and turn **ON** Plot1. For **Type**, select the last icon on the second line. For **Data List,** press 2nd [L1]. For **Data Axis**, select **X**. For **Mark**, select the first icon. Press ZOOM and 9 to display the normal plot.

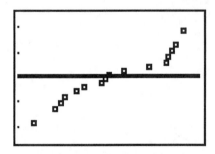

Data that is normally distributed will plot as a straight line. Although this graph is not perfectly straight, it is fairly linear and it is therefore reasonable to conclude that the data is approximately normal.

Repeat this process to check the other datasets for normality.

Exercise 4: Test the hypotheses: $H_o : \mu_1 = \mu_2 = \mu_3$ vs. H_a :at least one mean is different from the others. Press STAT, highlight **TESTS** and select **F:ANOVA(** and press ENTER, 2nd [L1] , 2nd [L2] , 2nd [L3] . Press ENTER and the results will be displayed on the screen.

Exercise 5: Repeat Exercises 1 -4 using this second set of data.